Moiré Fringes
in Strain Analysis

Moiré Fringes
in Strain Analysis

BY

PERICLES S. THEOCARIS, D.Sc., D.Appl.Sc.

Professor of Applied Mechanics and Director, Laboratory for Testing Materials, The National Technical University, Athens, Greece

THE QUEEN'S AWARD
TO INDUSTRY 1966

PERGAMON PRESS

OXFORD · LONDON · EDINBURGH · NEW YORK

TORONTO · SYDNEY · PARIS · BRAUNSCHWEIG

Pergamon Press Ltd., Headington Hill Hall, Oxford
4 & 5 Fitzroy Square, London W.1

Pergamon Press (Scotland) Ltd., 2 & 3 Teviot Place, Edinburgh 1

Pergamon Press Inc., Maxwell House, Fairview Park, Elmsford,
New York 10523

Pergamon of Canada Ltd., 207 Queen's Quay West, Toronto 1

Pergamon Press (Aust.) Pty. Ltd., 19a Boundary Street, Rushcutters Bay,
N.S.W. 2011, Australia

Pergamon Press S.A.R.L., 24 rue des Écoles, Paris 5e

Vieweg & Sohn GmbH, Burgplatz 1, Braunschweig

Printed in Great Britain by Thomas Nelson (Printers) Ltd., London & Edinburgh

To the memory of my mother

Contents

Preface

As EARLY as 1859 Foucault[1] proposed a method for testing lenses and optical systems by using low-frequency type gratings. In his famous publication Foucault suggested three different methods for testing optical systems, the second of which was based on gratings. Unfortunately, he did not develop this method because he considered it as less sensitive than the knife-edge method. This erroneous judgement was derived from the fact that he studied the phenomenon exclusively in the light of geometric optics, since he used gratings of very low frequency.

The phenomenon of moiré fringes was first described by Lord Rayleigh in 1874. In a paper "On the manufacture and theory of diffraction gratings"[2] he wrote:

> If two photograph copies containing the same number of lines to the inch be placed in contact, film to film, in such a manner that the lines are nearly parallel in the two gratings, a system of parallel bars develops itself, whose direction bisects the external angle between the directions of the original lines and whose distance increases as the angle of inclination diminishes. . . . When parallelism is very closely approached the bars become irregular in consequence of the imperfection of the rulings. This phenomenon might be made useful as a test.

Righi,[3] in 1887, extensively dealt with the distribution of light in figures formed by the superposed bars. He considered the case where the width of the bars as well as the transmittance of the gratings were variable. Moreover, the problem of the moiré fringe formation by combinations of overlapping circular and radial gratings with different centre-to-centre distances of the grating was considered. Finally, the creation of moiré fringe patterns by parallel line gratings and their application to the measurement of the relative displacement of the gratings was treated.

[1] Foucault, L., *Annls. Obs. Paris* **5**, 197 (1859).
[2] Lord Rayleigh (Strutt, J. W.), *Phil. Mag.* **47** (81) 193 (1874).
[3] Righi, A., *Nuovo Cim.* **21**, 203 (1887); **22**, 10 (1888).

The idea of Foucault found an application by Ronchi,[4] in 1922, for the optical testing of lenses and mirrors. A point source illuminated the optical system under test, which, when viewed through a low-frequency grating placed at the vicinity of the image of the light source, formed fringes. The moiré fringes of this arrangement were affected by any aberration of the optical system. Thus, this simple apparatus formed a sensitive achromatic interferometer.

Ronchi,[5] in 1925, studied the case of the moiré patterns created by the superposition of a line and a circular grating. It was shown that the form of the moiré fringes formed by the two overlapping gratings depended on the pitches of the gratings and that the resulting fringes are hyperbolas, parabolas or ellipses.

Raman and Datta[6] treated the phenomenon and gave the parametric equations of the fringes formed by the superposition of two zone gratings with different centre-to-centre distances. The moiré pattern formed consisted of parallel line fringes (Brewster bands) or families of circles.

The phenomena observed by Foucault, Lord Rayleigh, Righi, Ronchi, Raman and Datta did not enjoy the consideration which they merited until quite recently, despite the many advantages which they offered in metrology. This was due to the many difficulties encountered in the reproduction of satisfactory gratings and in the application of the methods.

The observations by Lord Rayleigh and Righi waited much longer than Foucault's ideas to be used as a measuring device. Tolenaar,[7] in 1945, seems to be the first who gave an interpretation of the moiré phenomenon based on geometric optics. He described the characteristic properties of moiré fringes formed by coarse gratings of equal or slightly different pitch in terms of their relative rigid-body translation, angular displacement or deformation. In

[4] Ronchi, V., *Riv. Ottica Mecc. Precis.* **2**, 19 (1922).

[5] Ronchi, V., La prova dei sistemi ottici, *Attual. Scient.*, No. 37 (N. Zanichelli, Bologna, 1925), Ch. 9.

[6] Raman, C. V., and Datta, S. K., *Trans. Opt. Soc. London* **27** (7) 51 (1926).

[7] Tolenaar, D., *Moiré interferentieverschijnselen bij rasterdruck*, Amsterdam, Institut voor Graphische Technik, 1945.

1952 Kaczer and Kroupa[8] applied the properties of moiré fringes to determine the strain components in a two-dimensional strain field.

Meanwhile gratings of good quality were designed for metrological applications and produced at a moderate cost by applying a novel principle introduced by Sir Thomas Merton[9] in England, for the manufacture of diffraction gratings by a reproduction from a turned master grating. The new methods of manufacture of diffraction and coarse gratings of high quality made their use very efficient for measurement and created new fields of application of moiré fringes in measuring processes. Indeed, in the last two decades a great number of new measuring applications have been developed. These techniques may be classified into two main categories, i.e. techniques for measurement of rigid body translations and angular displacement, which find their way to applications in automatic control or monitoring of programmed machining operations, and techniques for measurement of linear and angular displacements in deformed bodies.

In a paper for the *Applied Mechanics Reviews* in 1962 the author[10] reviewed 85 references covering the up-to-date most important publications related to applications of moiré fringes in metrology. The paper has been recently revised in order to be included in the *Applied Mechanics Surveys*,[11] a book published by the American Society of Mechanical Engineers in 1966 and containing all the main reviews published in *Applied Mechanics Reviews* from 1950 to 1965. The revised form of the review paper now included the papers published during the period 1961 to 1965 and totals 151 references. This difference shows clearly the important role which moiré fringes actually play in metrology.

The purpose of this book is to give a comprehensive description of the whole spectrum of methods and techniques using the moiré

8 Kaczer, J., and Kroupa, F., *Czech. J. Appl. Phys.* **1** (2) 80 (1952).

9 Merton, T., *J. Phys. Radium* **13** (2) 49 (1952).

10 Theocaris, P. S., *Appl. Mech. Rev.* **15** (5) 333 (1962).

11 Theocaris, P. S., *Applied Mechanics Surveys*, Spartan Books, Washington, D.C., 1966, p. 613.

fringe phenomenon for the measurement of strains in deformed bodies and in engineering structures. Each method is developed in detail in order to point out its dependence on certain physical principles as well as its eventual connection and relationship with other similar methods. Examples of applications of each technique to particular metrological problems were added in order to clarify further the possibilities of each method as a measuring device.

An attempt is made to give the simplest and most practical presentation of the various methods and to make their approach straightforward and simple by avoiding the introduction of mathematical theories and generalizations in cases where they are not needed.

This treatment should enable researchers as well as newcomers in the field to effectively solve many of their problems by choosing the most effective and reliable method and avoid the embarrassment and sometimes the eventual impasse produced by complicated mathematical theories developed to describe complex optical phenomena which have been transplanted to simple cases because of a certain boasting mood of their inventors.

Also for simplicity, the development of the various methods follows a simple pattern, which has the advantage of introducing a few new ideas at a time. It is the author's hode that the book will be used as a basis for the future researcher and student to develop his own new and more fruitful ideas and methods.

Extended parts of this book are mainly based on research work carried out in the Laboratory for Testing Materials of the Athens National Technical University. It is a pleasure to record my appreciation for the co-operation of the technicians of the laboratory for the preparation of the experiments, of Mr. A. Koutsambessis for his valuable help in many of the experiments described in the book and for preparing the photographic illustrations of the text, of Miss E. H. Lekka for typing the manuscript and tracing the ink-drawings, and, finally, of Dr. A. Vafiadakis for reading the manuscript and for his efficient help.

CHAPTER 1

General Theory

1.1. Introduction

Interference fringes between two beams of monochromatic light have long been used for the most refined measurement of length, including the establishment of fundamental length standards in terms of the wavelength of light. In all these measurements, length is determined in terms of the wavelength used, which is approximately one fifty-thousandth of an inch. As the light path in the systems employed is doubled by reflection the passage of one fringe past a point is produced by a relative movement of half a wavelength, i.e. one hundred-thousandth of an inch. It is only rarely that precision of this order is required in current laboratory work. Since laboratory measurements usually require less precision, interference methods would only be convenient if they could be adapted to wavelengths many times greater (up to fifty) than those of visible light. But, since there are no visible emission lines greater than two or three times that of the green light, the use of ordinary interference methods for measurement purposes is impractical.

However, it is possible to produce fringes with white light by using two coarse slit-and-bar gratings. These consist of glass plates on which opaque bars are ruled at regular intervals leaving transparent slits usually of equal width, thus forming an amplitude grating of a 50 per cent transmittance. If two amplitude gratings of approximately equal frequency are face to face superposed with their rulings nearly parallel and viewed in a diffused light background, sharp moiré fringes will be observed. No special illuminating and observing system is necessary for observing the moiré

1

fringes provided that the two gratings are close to each other. The moiré fringes formed are similar to the fringes formed by the superposition of two diffraction gratings. Satisfactory observation of moiré fringes without special optical equipment is only possible if the gratings are of comparatively low frequency, preferably not exceeding 1000 lines per inch. Gratings with pitches in this range are the process screens used by the printers for half-tone reproduction. They can be reproduced by a photographic process from a master screen ruled on special machines. A photographic reproduction without significantly affecting the balanced transmittance of the grating can be achieved with special photographic films and plates if the frequency of the master remains in the limit of 1000 lines per inch.

The amplitude gratings under the usual conditions of observation do not exhibit features shown with finer rulings, especially with transparent diffraction gratings. Diffraction gratings are normally ruled on glass plates or films and consist of families of parallel furrows forming curves of first, second or higher degree. Usually diffraction gratings have straight-line furrows in frequencies of the order of several thousand lines per inch. The transmittance of the grating remains the same as the transmittance of the flat plate or film since the furrows do not obstruct the transmission of light in any useful sense of the term. The moiré fringes formed by diffraction gratings are therefore due to diffraction and interference of diffracted light beams emerging from each furrow. On the contrary, the explanation which is put forward for the moiré fringes formed by amplitude gratings, attributes the fringes to the obstruction of light by the opaque bars, the obstruction being greater where the bars of one grating cover the transparent slits of the other and least where the two sets of bars coincide or intersect. This explanation is based on rectilinear propagation of light and is only approximate. It is satisfactory for large macroscopic objects, but becomes less satisfactory with reduction of the width of the bars and fails entirely to account for optical behaviour of microscopic objects.

The explanation of formation of moiré fringes based on the

obstruction of light, assuming a rectilinear propagation of light, does not exclude the simultaneous existence of diffraction and optical interference phenomena. In reality obstruction phenomena

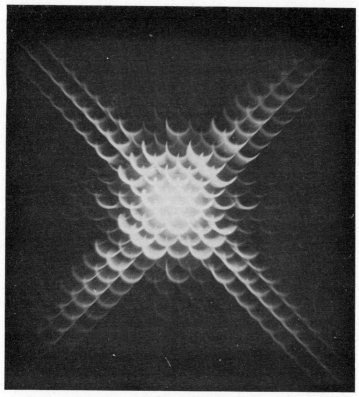

FIG. 1.1. Diffracted image of the sun viewed at Athens a few minutes before its total eclipse of 20 May 1966 through a filter and an amplitude grating of a frequency of 1000 lines per inch.

and diffraction and interference phenomena coexist, and the predominance of the one phenomenon over the other depends on the type of illumination besides the type of the grating.

If two similar gratings are face to face superposed and illuminated by a strongly coherent monochromatic point source, the angular extent of which is restricted by a diaphragm, the moiré fringes formed may be attributed to the obstructed rectilinearly propagated light if the distance between the gratings is infinitesimal and their frequency low. The interference phenomena due to the diffraction of light are not negligible but are superposed on the fringe patterns formed by rectilinear propagation.

Figure 1.1 shows the sun viewed at Athens a few minutes before its total eclipse of 20 May 1966 through a filter and an amplitude grating of a frequency of 1000 lines per inch. Since the sun illuminated the photographic plate by strongly coherent light, diffraction phenomena were dominant with the amplitude type of grating. If the same grating was illuminated by a non-coherent source produced by a light diffuser placed in the proximity of the grating, interference of this grating with a similar grating will form moiré fringes solely due to the obstruction of light by the opaque bars. In this arrangement, optical interference and diffraction phenomena are insignificant. Therefore, this case corresponds to the one extremity where the totality of the phenomenon of formation of moiré fringes is due to changes in the intensity distributions of light. But this intensity variation may be considered merely an analogous phenomenon to optical interference, where the shadowed regions of the moiré pattern due to obstruction of light correspond to the lowest intensity regions of an interferogram. That is perhaps an explanation and a justification for the term *mechanical interference* given to the formation of the moiré fringes by the obstruction of light. The term, although generally accepted, is not the appropriate one. Mechanical interference could be better called either figure or geometric interference.

Figure 1.2 shows a moiré pattern formed by two identical gratings relatively displaced by a small angle ϑ and illuminated by diffused light. The moiré pattern is solely due to the intensity variation of light passing successively the two gratings.

In the case where two furrowed gratings are illuminated by a light diffuser, no interference phenomena are produced. On the contrary, if a monochromator is used to illuminate these gratings

FIG. 1.2. Moiré pattern formed by two identical gratings angularly displaced by a small angle ϑ and illuminated by a diffuser.

(i.e. a strongly coherent monochromatic light source of restricted angular extent), moiré fringes appear solely due to diffraction and optical interference phenomena.

Since in almost all cases of application of interference phenomena to strain analysis amplitude gratings are used of a frequency lower than 1000 lines per inch, the phenomenon of production of moiré fringes by the obstruction theory will be described in detail. The treatment which follows is mainly confined to the properties of moiré fringes formed by amplitude-type gratings under a non-coherent type of illumination. However, there are cases where coherently illuminated diffraction gratings are used in applications of moiré fringes in strain analysis. These cases will be described separately using optical interference principles.

If two superposed line gratings are viewed in a bright background with their rulings exactly parallel, the field appears having a uniform brightness. If the rulings of the one grating are angularly displaced to disturb their parallelism, the gratings are transversed by a number of straight equidistant fringes the pitch of which decreases as the acute angle of relative angular displacement increases. Thus the fringe pitch can attain any desired value by adjusting the acute angle formed between the rulings. If the gratings are identical the fringes run perpendicular to the bisector of the interruling angle.

In the case of initially identical gratings with no angular displacement, moiré fringes are formed if the pitch of one of the gratings is changed due to an externally applied deformation. If the applied deformation is in the direction of the normal to the rulings, the moiré fringes are parallel to the rulings. If the deformation is parallel to the direction of the rulings no moiré fringes are formed. A deformation applied in an oblique direction to the rulings can be analysed into two translation components, one normal and the other parallel to the rulings. While the normal component forms moiré fringes, the parallel component contributes nothing to the moiré pattern.

It is worth while pointing out the difference between the moiré patterns formed by relative angular displacement between two identical gratings and those formed by an initial relative disparity in pitch between the two gratings. The moiré fringes formed by angular displacement are deviating from the normal to the rulings of the one grating by an angle equal to the half of angular displacement. Accordingly, the displacements associated with the pure angular displacement present a weak gradient along the axis normal to the rulings but a steep gradient along its perpendicular axis. The moiré fringes formed by the superposition of two initially mismatched gratings are straight, parallel bands uniformly spaced and parallel to the rulings of the gratings.

Therefore, while the disparity fringes are parallel to the rulings, the moiré fringes formed by a small angular displacement are oriented essentially in a perpendicular direction.

The formation of moiré fringes due to disparity in pitch of the two gratings was used by Sir Thomas Merton[1] in 1950 to test gratings for an eventual presence of periodic errors. For this purpose two replica copies of the grating under test were superposed and viewed in the appropriate type of light. It was only in the very rare case when the defect of the one replica exactly coincided with the defect of the other replica that it was impossible to detect the existing defects. Generally, defects are not coincident, and each grating acts as magnificator of the other showing all the eventually existing defects. Figure 1.3 shows the magnification of a 1000-line grating and its defect by superposition of a similar grating with the same defect. The vertical line of the figure corresponds to a ply of the original grating.

The same principle is actually used in electron microscopy to detect dislocations. Pashley, Menter and Basset[2,3] showed that atomic dislocations can be detected when two appropriately oriented thin lamellae showing their crystal lattice are superposed and viewed in an electron microscope. If the lattice periodicities of the two lamellae are parallel, the moiré pattern formed shows the

[1] Merton, T., *Proc. Roy. Soc. London* A **201**, 187 (1950).

[2] Pashley, D. W., Menter, J. W., and Basset, G. A., *Nature* **179**, 752 (1957).

[3] Basset, G. A., Menter, J. W., and Pashley, D. W., *Proc. Roy. Soc. London* A **246**, 345 (1958).

eventual dislocations appearing as extra half lines. Similar results can be obtained by relative angular displacement of the lamellae. As in the case of rulings, the authors have pointed out that it is only when a dislocation in one lattice is superposed on the dislocation in the other that the dislocation is invisible in the moiré pattern.

Oblique illumination and viewing of amplitude gratings is used to test the overall flatness of irregular matt or specularly polished

FIG. 1.3. Magnification of a line grating of an original frequency of 1000 lines per inch and its defect by super-posing it on a similar grating.

flat or curved surfaces. In this case the shadow or reflection of the grating is deformed and distorted by its projection on to the irregular surface and, when these interfere with the original grating, form moiré patterns which are a measure of magnified surface irregularities, i.e. change of height or slope.

Another characteristic phenomenon of moiré fringes, which is used for the measurement of displacement, is produced when one grating is moved in its own plane while the other remains fixed in relation to the observer. The fringes move across the field in the direction which is normal to the rulings and the number of fringes which pass any point of the field is equal to the number of rulings of the moving grating which passes the same point. If each grating is rigidly attached to the relatively moving components of a structure it is possible to measure the relative displacement of the two components in terms of the pitch of rulings by counting the number of fringes passing any fixed point of one grating or the brightness cycles at the same point.

Fringes are also formed if the pitches of the gratings are to a close approximation in the ratio of two small integers, for example 2:1, 3:1, 3:2. In addition, fringes are as well formed by projecting an image of one grating on to the surface of another, in which case gratings of any relative pitch may be used. In this case the image of the one grating is properly magnified to form a projection of the grating of equal pitch to the pitch of the second grating. It is also possible to project the image of a grating on a reflecting grating and view the set obliquely. Another combination is to use a transparent and a reflecting grating in contact or in close proximity to each other. In all these cases the illumination must originate from a light source placed in front of the reflecting grating so that the light beam passes twice through the transparent grating once on its incidence to and again on its reflection from the reflecting grating.

The changes produced in the separation, direction and position of fringes are relatively large, compared with the respective deformations or movements of the grating causing these changes. Therefore, *the behaviour of two superposed gratings provides highly sensitive means of measuring small linear and angular displacements or expansions of gratings*.

The phenomenon of formation of moiré fringes from the irregularities of the one grating due to its deformation is extensively used in a variety of ways for the measurement of displacements in plane stress and plane strain problems or deflections in flexed structures. The techniques related to the measurements of displacements and deflections will be classified and described in the following in a rational order starting with methods capable of recording small in-plane displacements of deformable bodies and followed by methods capable of measuring large lateral deflections, slopes and curvatures.

The displacement of moiré fringes due to a rigid body translation of the one grating relative to its pair has found extensive application in automatic control of programmed operations. This type of application of moiré fringes is beyond the scope of this book.

1.2. Nature of the moiré phenomenon

Before a discussion on the moiré phenomenon is undertaken it is necessary to introduce some of the essential terms which are used in the text.

In the particular problems and applications of moiré fringes to metrology only perfect gratings will be considered in the development of the theory explaining each phenomenon. A perfect grating, viewed under a microscope, appears to consist of parallel equi-spaced opaque bars of constant width separated by transparent slits of equal width. The parallel bars may form any family of first, second or higher degree curves. The amplitude type of grating has some characteristic features which are repeated in each bar.

The ensemble made by an opaque bar or groove and its adjacent transparent slit is termed a *ruling* or a *line* of the grating. The distance between the edge of a generic opaque bar or groove and the corresponding edge of the next bar or groove is termed the *pitch* of the grating and is denoted by p. Usually gratings are specified by frequency, which is the reciprocal of the pitch and is expressed by the number of bars or grooves per unit length.

The gratings used in strain analysis applications are considered as having a 50 per cent transmittance, that is having opaque bars or grooves and transparent slits of equal width.

A direction perpendicular to the rulings in the plane of the grating is called the *principal direction*, while a direction parallel to the rulings is called the *secondary direction*. The distance between two successive moiré fringes formed by the superposition of two similar gratings is called the *interfringe spacing* and is denoted by f.

This distance may be measured between either the two darkest or two brightest centre lines of the successive moiré fringes.

If the rulings of each interacting grating are regarded as indexed families of curves, the moiré pattern formed by the interference of the two gratings is more distinct at the points of intersections, the indices of which satisfy some simple relation. Let $R(x,y) = k$ express the family of curves corresponding to the grating, which remains constant during measurement and is called the *reference grating* (*RG*), and $S(x,y) = l$ express the family of curves corresponding to the deformed grating, which is called the *specimen grating* (*SG*); k and l are the indexing parameters running over a subset of the real integers. The values of the real integers k and l, which define the spacing of the individual curves, may be positive or negative and measured by a unit which renders the grating sufficiently dense; x and y are the co-ordinates of any point in the plane of figure, which is the plane of superposition of the gratings. The functions $R(x,y)$ and $S(x,y)$ define the form of the individual curves of the families. The resultant moiré pattern is an indexed family of curves $M(x,y) = m$ with an index m, which runs over some subset of real integers. The moiré index m must satisfy an indicial equation which, for the totality of two-grating simple systems, takes the form

$$k \pm l = m. \tag{1.1}$$

If two gratings, expressed by the indicial equations $R(x,y) = k$ and $S(x,y) = l$, are superposed so that their curves form a variable angle $\vartheta(x,y)$, the curvilinear quadrangles have their corners lying on a bright moiré fringe. Figure 1.4 shows that the two curvilinear diagonals of each quadrangle correspond to moiré fringes satisfying either the equation $k+l = m$ or the equation $k-l = m$. Points A, B, C, D, ..., correspond to a moiré fringe for which $(k+l) =$ const., while points E, F, G, H, ..., correspond to a moiré fringe for which $(k-l) =$ const.

The family of moiré fringes for which the equation $k-l = m$ is valid is called the *subtractive moiré pattern*, while the family of moiré fringes for which the equation $k+l = m$ is valid is called the *additive moiré pattern*. The *effective or visible moiré pattern* is the pattern in which the interfringe spacing is the longest, that is the pattern in which the moiré fringes coincide with the shortest diagonals of the individual quadrangles. The effective moiré pattern in Figure 1.4 is the pattern containing the fringe *EFGH* and belongs to a subtractive moiré pattern.

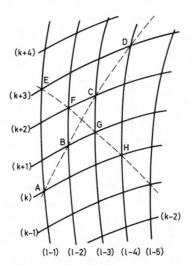

Fig. 1.4. Schematic representation of indicial formation of subtractive and additive moiré patterns.

Since the shape and dimensions of the individual quadrangles change with the relative displacement of the two gratings, it is possible that the effective moiré pattern commutes from the subtractive type to the additive type and vice versa. By doing so, the moiré pattern traverses the so-called *commutation moiré boundary* or, simply, the *moiré boundary*. This boundary may be defined as the region which contains all individual rectangles or squares.

In order to study the nature of the moiré phenomenon it suffices to consider the interference of two line gratings, the rulings of which form an acute angle ϑ. The nature of more complicated cases may be derived by inductive reasoning. Figure 1.5 shows two identical gratings of pitch p and 50 per cent transmittance enlarged sufficiently to display the details of the rulings.

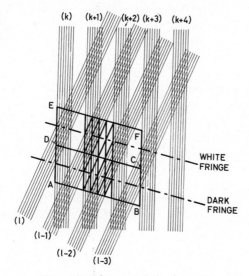

FIG. 1.5. Diagram of the intensity distribution across moiré figures.

The superposition of the gratings produces rhombuses of three different shades of darkness, i.e. white, simple black and double black rhombuses. The shortest diagonals of the elementary simple black rhombuses form straight lines which are the centre lines of the dark moiré fringes, while the shortest diagonals of the elementary successive white and double black rhombuses form the centre lines of the bright moiré fringes. The same construction can be repeated for the longest diagonals of the rhombuses. In this case the additive moiré pattern is formed.

From zones $ABCD$ and $CDEF$ in Figure 1.5, which belong to a dark and its successive white moiré fringe, it can be derived that each dark moiré fringe contains six parts of simple black shade, while the remaining two parts are equally divided between a white and a double black shade. Thus, in a dark moiré fringe 87·5 per cent is black of which 12·5 per cent is double black and the remaining 12·5 per cent is white. By repeating the procedure for the adjacent white fringe, it can be found that three parts belong to a double black shade, another three parts to a white shade and the remaining two parts are simple black. If, in a black fringe four simple black parts are changed, two to white and, for balance, the other two to double black, the adjacent white fringe is formed. In percentage proportions 62·5 per cent of a white fringe is covered by black areas from which a 37·5 per cent is double black and the remaining 37·5 per cent is white. Thus, in a dark or bright moiré fringe all regions are neither black nor bright. The separation of successive dark and bright fringes is achieved by tripling the white areas of the dark fringes.

In the general case where the rulings of the gratings follow curves of any degree the procedure of formation of moiré fringes follows the same principle and the rhombuses become curvilinear quadrangles, the shortest diagonal of which yields the moiré fringes.

1.3. Mathematical analysis of moiré fringes

Consider two superposed families of curves forming a moiré pattern. Let the two families be expressed by the indicial equations

$$S(x,y) = k \quad \text{and} \quad R(x,y) = l. \tag{1.2}$$

By tracing the diagonals of rhomboids along the points of intersection of the curves two new families of curves are formed which are denoted by $M_a(x,y) = m_a$ and $M_s(x,y) = m_s$. The families of curves may be considered as the rulings of the two gratings while the families of curves formed by the diagonals of the intersected rulings are the additive (M_a) and subtractive (M_s) moiré patterns.

The distances between successive lines for the two original families of curves express their pitches and for the families of diagonals express the interfringe spacings of the two moiré patterns. Figure 1.6 shows the subtractive moiré fringes formed by two families of curves (k) and (l). The two successive moiré fringes are indexed as (m) and $(m+1)$. The interfringe spacing f_{ms} is defined by the distance OK on the normal n at O. From the triangle $(OO'K)$ the interfringe spacing f_{ms} is given by

$$f_{ms} = a \sin(\vartheta - \varphi), \tag{1.3}$$

where a is given by

$$a = (x^2_{,m} + y^2_{,m})^{\frac{1}{2}}, \tag{1.4}$$

where commas indicate the first partial derivatives of the corresponding functions with respect to the subscripts. In this relation it was taken into consideration that $dm = 1$, that is, the difference in indices between two consecutive moiré fringes is equal to unity.

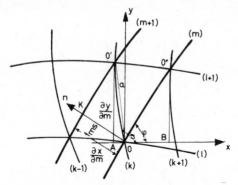

FIG. 1.6. Geometry of formation of moiré fringes.

Angles ϑ and φ may be readily defined from the geometry of Figure 1.6 and can be expressed by

$$\tan \vartheta = \frac{y_{,m}}{x_{,m}} \tag{1.5}$$

and

$$\tan \varphi = -\frac{M_{,x}}{M_{,y}}. \tag{1.6}$$

Eliminating ϑ and φ from relation (1.3) and taking into consideration that

$$\frac{dM}{dm} = M_{,x}\, x_{,m} + M_{,y}\, y_{,m} = \frac{\partial m}{\partial m} = 1$$

it can be shown that

$$f_{ms} = (M^2{}_{,x} + M^2{}_{,y})^{-\frac{1}{2}}. \tag{1.7}$$

Since for the subtractive moiré fringes

$$S(x,y) - R(x,y) = M(x,y), \tag{1.8}$$

then relation (1.7) becomes

$$f_{ms} = [(S_{,x} - R_{,x})^2 + (S_{,y} - R_{,y})^2]^{-\frac{1}{2}}. \tag{1.9}$$

For the additive moiré fringes, where the indicial equation takes the form

$$S(x,y) + R(x,y) = M(x,y), \tag{1.10}$$

the interfringe spacing f_{ma} is given by

$$f_{ma} = [(S_{,x} + R_{,x})^2 + (S_{,y} + R_{,y})^2]^{-\frac{1}{2}}. \tag{1.11}$$

By a similar reasoning, though much simpler, it can be shown that for the line pitches p_r and p_s of the two gratings

$$p_r = (R^2{}_{,x} + R^2{}_{,y})^{-\frac{1}{2}}, \tag{1.12}$$

$$p_s = (S^2{}_{,x} + S^2{}_{,y})^{-\frac{1}{2}}. \tag{1.13}$$

The equality of the interfringe spacings for the subtractive and the additive moiré patterns expresses the parametric equation for the commutation moiré boundary, that is

$$f_{ms} = f_{ma}. \tag{1.14}$$

Substituting the values for f_{ms} and f_{ma} from eqns. (1.9) and (1.11) it can be deduced that

$$\psi(x,\, y) \equiv R_{,x}\, S_{,x} + R_{,y}\, S_{,y} = 0. \tag{1.15}$$

The condition for the subtractive moiré pattern to be effective is

$$f_{ms} > f_{ma}.$$

Therefore $\psi(x,y) \equiv R_{,x} S_{,x} + R_{,y} S_{,y} > 0.$ (1.16)

The subtractive moiré pattern is effective in regions of the moiré field where the sum of the products of the first partial derivatives of the functions expressing the families of rulings with respect to the principal axes of the gratings is positive. In the regions where this sum is negative, the additive moiré pattern is effective, and it will appear in the moiré pattern.

Introducing relations (1.12) and (1.13) into relations (1.9) and (1.11) the interfringe spacings f_{ms} and f_{ma} are given by

$$f_m = \frac{p_r p_s}{(p_r^2 + p_s^2 \pm 2\psi p_r^2 p_s^2)^{\frac{1}{2}}},$$ (1.17)

where the positive sign corresponds to the additive moiré pattern and the negative sign corresponds to the subtractive moiré pattern.

Using relations (1.12), (1.13) and (1.16) it can be shown that

$$\psi p_r p_s = \frac{1}{\sqrt{\left[1 + \left(\dfrac{R_{,y}S_{,x} - R_{,x}S_{,y}}{R_{,x}S_{,x} + R_{,y}S_{,y}}\right)^2\right]}} = \frac{1}{\sqrt{(1 + \tan^2 \vartheta)}} = \cos \vartheta$$ (1.18)

and therefore relation (1.19) becomes

$$f_m = \frac{p_r p_s}{(p_r^2 + p_s^2 \pm 2 p_r p_s \cos \vartheta)^{\frac{1}{2}}}.$$ (1.19)

Equation (1.17) indicates that the interfringe spacing along the commutation moiré boundary where $\psi = 0$ is given by

$$_{J mc} = \frac{p_r p_s}{\sqrt{(p_r^2 + p_s^2)}}.$$ (1.20)

Since for the subtractive moiré fringes, where ψ is positive, the negative sign is valid in eqn. (1.17), while for the additive moiré fringes, where ψ is negative, the positive sign is valid in the same equation, it can be concluded that a positive quantity is always subtracted from the sum $(p_r^2 + p_s^2)$ in the denominator of eqn. (1.17) and therefore

$$f_m > f_{mc} = \frac{p_r p_s}{\sqrt{(p_r{}^2 + p_s{}^2)}}.$$ (1.21)

It is worth while mentioning that relation (1.19) may be readily established from the geometry of the parallelogram formed by every two adjacent rulings of each grating, where ϑ expresses the acute angle between the principal directions of the rulings.

CHAPTER 2

Moiré Patterns formed by Line Gratings

2.1. Indicial representation of moiré patterns

Lehmann and Wiemer,[1] as early as 1953, presented the theory of moiré patterns formed by different types of gratings based on the indicial representation method. Pirard,[2] in 1960, gave a complete discussion of moiré phenomena produced by line, radial, circular and parabolic gratings, as well as their most interesting combinations which was also based on the method of indicial representation of sets of curves. Oster, Wasserman and Zwerling[3] used the indicial approach, which was generalized to the study of various types of moiré patterns, and applied the moiré technique to measurements of refractive index gradients as well as to studies of the optical properties of lenses. Moreover, they pointed out the analogy of moiré to hydrodynamics, electrostatics, physical optics, etc.

Besides the above method, several other techniques have been developed for the representation of moiré patterns. These are based on elementary or differential geometry as well as on vectorial and tensorial concepts. While these methods may yield simple and straightforward solutions in particular combinations of gratings, the indicial representation method has the advantage of yielding a unified manner of solving all cases of combinations of gratings which is simple and avoids the complicated mathematics which are

[1] Lehmann, R., and Wiemer, A., *Feingeräte Technik* **2** (5) 199 (1953).

[2] Pirard, A., *Analyse des Contraintes*, *Mém. GAMAC* **5** (2) 1 (1960).

[3] Oster, G., Wasserman, M., and Zwerling, C., *J. Opt. Soc. Am.* **54** (2) 169 (1964).

necessary with the other methods. For an extensive review of the
various methods of solutions of moiré patterns formed by any two
sets of gratings, see ref. 2 on p. xi of the Preface.

Let two line gratings be face to face superposed with their rulings
subtending an acute angle ϑ, the pitch of the specimen grating

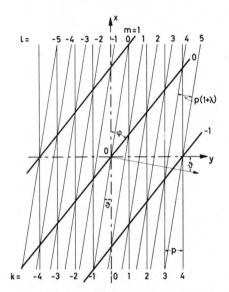

FIG. 2.1. Schematic representation
of indicial formation of a subtractive moiré pattern by two
identical gratings angularly displaced through an angle ϑ.

$S(x,y)=k$ be equal to p, while the pitch of the reference grating
$R(x,y) = l$ is equal to $p(1 + \lambda)$, where λ is an infinitesimal quantity
positive or negative. Both families of straight lines in the gratings
are indexed by k and l, which run from zero to plus or minus infinity. A system of Cartesian coordinates is referred to both gratings with its axes coinciding with line $k = 0$ and its normal passing
through the intersection of lines $k = 0$ and $l = 0$ (Fig. 2.1).

The equation of the family of lines in SG is given by

$$S(x,y) \equiv y = pk, \qquad (2.1)$$

and the equation for the second family of lines in RG is expressed by

$$[y \cot \vartheta - lp(1+\lambda)/\sin \vartheta] = x. \qquad (2.2)$$

Equation (2.2), solved with respect to l, gives

$$R(x,y) \equiv (y \cos \vartheta - x \sin \vartheta) = lp(1+\lambda). \qquad (2.3)$$

The effective moiré fringes run along the shortest diagonals of the parallelograms formed by the intersections of the rulings. Since, in this case, all parallelograms are equal the effective moiré is the subtractive moiré since $f_{ms} > f_{ma}$.

Substituting in eqn. (1.10) the values of the derivatives $R_{,x}, R_{,y}$ and $S_{,x}, S_{,y}$ it is deduced that

$$\psi(x,y) = \frac{\cos \vartheta}{p^2(1+\lambda)}. \qquad (2.4)$$

This equation becomes equal to zero only in the case where $\vartheta = 90°$, i.e. the rulings of the gratings become orthogonal and no moiré pattern is formed. Equation (2.4) remains always positive for acute angles ϑ and therefore the subtractive moiré pattern is effective everywhere.

Since the subtractive moiré pattern is valid throughout the field, the indicial equation for the moiré pattern is given by eqn. (1.1) with the negative sign, i.e.

$$(k-l) = m. \qquad (2.5)$$

The equation for the moiré fringes is also given by

$$\left(y \cot \varphi + \frac{mf}{\sin \varphi} \right) = x, \qquad (2.6)$$

where φ is the angle between the lines $y = kp$ and the moiré fringes and f the interfringe spacing of the effective moiré pattern.

Eliminating the indices k and l from relations (2.1), (2.3) and

(2.5) yields the following equation for the moiré fringes:

$$y\left[\frac{\cos \vartheta - (1+\lambda)}{\sin \vartheta}\right] + \frac{mp(1+\lambda)}{\sin \vartheta} = x. \qquad (2.7)$$

Equations (2.6) and (2.7) are identical. Therefore, by equating coefficients it is deduced that

$$f = p(1+\lambda).\left[\lambda^2 \cos^2 \frac{\vartheta}{2} + (2+\lambda)^2 \sin^2 \frac{\vartheta}{2}\right]^{-\frac{1}{2}}, \qquad (2.8)$$

$$\sin \varphi = \sin \vartheta \left[\lambda^2 \cos^2 \frac{\vartheta}{2} + (2+\lambda)^2 \sin^2 \frac{\vartheta}{2}\right]^{-\frac{1}{2}}. \qquad (2.9)$$

If the gratings are identical, i.e. $\lambda = 0$, relations (2.8) and (2.9) yield

$$f = \frac{p}{2 \sin \vartheta/2}, \qquad (2.10)$$

and

$$\sin \varphi = \cos \frac{\vartheta}{2}. \qquad (2.11)$$

It can be readily deduced from eqn. (2.11) that the moiré fringes are bisecting the supplementary angle of angle ϑ. The same result can be derived by considering that the elementary parallelogram formed by the intersections of the two sets of rulings is reduced for $\lambda = 0$ to a rhombus, the diagonals of which bisect the angles of the corners.

If the gratings are parallel, i.e. $\vartheta = 0$, relations (2.8) and (2.9) reduce to

$$f = \frac{p(1+\lambda)}{\lambda} \qquad (2.12)$$

and

$$\sin \varphi = 0. \qquad (2.13)$$

The moiré fringes formed by parallel gratings of different pitch are parallel to the rulings of the gratings (Fig. 2.2). Relations (2.8) to (2.13) are exact and valid for any value of λ and ϑ. But relations (2.10) and (2.12) imply that both quantities λ and ϑ must be small in order to obtain visible moiré patterns.

For $\lambda = 0$ and small values of angles ϑ, i.e. in the range where sines can be replaced by the angles, relations (2.10) and (2.11) yield

$$f = \frac{p}{\vartheta},$$ (2.14)

and $$\sin \varphi \approx 1.$$ (2.15)

FIG. 2.2. Moiré fringes formed by parallel gratings of slightly different pitch. The fringes run parallel to the rulings of the gratings.

The moiré fringes formed by a small angular displacement of identical gratings have an interfringe spacing f, which is equal to the pitch p divided by angle ϑ. These fringes are approximately normal to the rulings of the gratings.

If the pitches of the gratings are different and the acute angle ϑ

is small, the interfringe spacing f and the angle φ are given by the relations

$$f = \frac{p}{(\lambda^2 + \vartheta^2)^{\frac{1}{2}}} \tag{2.16}$$

and

$$\tan \varphi = \frac{\vartheta}{\lambda}. \tag{2.17}$$

If the angle ϑ equals zero, relation (2.16) is reduced to

$$f = \frac{p}{\lambda}, \tag{2.18}$$

which is approximately equal to eqn. (2.12). If $\lambda = 0$, relation (2.16) reduces to eqn. (2.14).

It is worth while mentioning the simple relation for angle ω formed between the bisectors of the acute angle ϑ and the moiré fringes. The expression for the tangent of this angle may be readily derived from eqn. (2.9) and expressed by

$$\tan \omega = \frac{\lambda + 2}{\lambda} \tan \frac{\vartheta}{2}. \tag{2.19}$$

Thus the tangent of the angle formed by the principal median plane and the moiré fringes is equal to the tangent of half the acute angle between the gratings multiplied by the ratio $(\lambda + 2)/\lambda$.

2.2. Measurement of the components of strain

Consider the general case of a two-dimensional specimen under conditions of plane stress or strain. Let u and v be the components of displacement in the x, y directions respectively. For small values of strain the components of strain $\varepsilon_x, \varepsilon_y, \gamma_{xy}$ are given by the relations

$$\varepsilon_x = \frac{\partial u}{\partial x}, \quad \varepsilon_y = \frac{\partial v}{y}, \quad \text{and} \quad \gamma_{xy} = \frac{\partial u}{\partial y} + \frac{\partial v}{\partial x}. \tag{2.20}$$

The theory developed in this section follows mainly an approach which has been described by Dantu.[4] A crossed grating, SG, is attached to the specimen before deformation with its principal and secondary directions parallel to the principal axes of the specimen. A similar grating, RG, reproduced on a photographic glass plate, is superposed on the deformed grating, SG, after deformation. The specimen grating follows the deformation of the body and each family of rulings follows the normal component of deformation parallel to its principal direction, while the angular displacement of the body is given by the respective displacements of the two families of rulings parallel to their secondary directions. If the one family of rulings is displaced by a displacement $u(x)$, normal to the rulings, and an angular displacement γ_{yy} yielding a displacement $u_r(y)$ parallel to the rulings, the total displacement $U(x,y)$ is given by the geometric sum

$$U(x,y) = u(x) + u_r(y).$$

The angular displacement due to the small angle γ_{yy} corresponds to a fictitious shift equal to

$$u_r(y) = \gamma_{yy} \cdot y.$$

Then
$$U(x,y) = u(x) + \gamma_{yy}y. \tag{2.21}$$

The normal strain ε_x and the one term γ_{yy} of the shear strain γ_{xy} are derived by differentiating eqn. (2.21) with respect to x and y, i.e.

$$\frac{\partial U}{\partial x} = \frac{\partial u}{\partial x} = \varepsilon_x \tag{2.22}$$

and
$$\frac{\partial U}{\partial y} = \gamma_{yy}. \tag{2.23}$$

Similarly, the total displacement $V(x,y)$ of a grating with the rulings initially parallel to the x-axis is given by

$$V(x,y) = v(y) + v_r(x) \tag{2.24}$$

[4] Dantu, P., Laboratoire Central des Ponts et Chaussées, Publ. No. 57–6 (1957).

and
$$\frac{\partial V}{\partial y} = \frac{\partial v}{\partial y} = \varepsilon_y, \tag{2.25}$$

$$\frac{\partial V}{\partial x} = \gamma_{xx}. \tag{2.26}$$

The shear strain γ_{xy} of the element of the body is given by adding the terms γ_{xx} and γ_{yy} given by eqns. (2.23) and (2.26):

$$\gamma_{xy} = \frac{\partial U}{\partial y} + \frac{\partial V}{\partial x}. \tag{2.27}$$

For the complete definition of the strain components in a two-dimensional strain field it is necessary to dispose two gratings with their rulings subtending any angle. It is expedient to use ortho-gonally crossed gratings, thus considerably reducing the amount of necessary calculations for the evaluation of the components of strains. While the specimen grating must be a crossed grating, the reference grating may be a line grating, which is superposed to *SG* twice consecutively with its rulings parallel to the two principal directions of *SG*. The superposition of a line grating on a crossed grating has the advantage of separately producing each family of moiré fringes, thus excluding the influence of the rigid-body angular displacement of the strain field caused by the deformation of the body compatible with the distribution of strain. The moiré fringes in this case are sharp and distinguishable from the family of moiré fringes formed by the orthogonal set of rulings and com-pletely exclude the interweaving of the two families of fringes, which results in a disguising of the identity of each family. More-over, the rounding of crossed moiré fringes developed at the inter-sections of the two families of fringes, which results in a consider-able increase of the width of each fringe, is avoided in this manner.

Another advantage of using line gratings as reference gratings is the flexibility in using gratings with different pitches along the two principal directions. This flexibility in freely choosing the suitable pitch of the grating results in moiré patterns with an optimum density of fringes and an optimum width of each fringe. These two factors considerably increase the accuracy of the method. However, the use of separate line gratings as reference gratings successively superposed on the crossed specimen grating results in a variable angle between the two sets of reference rulings. Relations (2.22) and (2.25) show that small deviations from the orthogonality of the two sets of rulings do not affect the values of normal components of strains. On the contrary, the shear components of strain are considerably affected. This drawback may be overcome by taking care to accurately orient the two positions of reference grating according to orientation marks traced along the principal axes of the specimen, which do not change their relative position during deformation.

In most cases where symmetrical specimens are studied with symmetrically applied loads, the symmetry of moiré patterns can be used as a sensitive indicator of the orientation of the reference grating. In these cases large errors introduced by the misorientation of *RG* can be avoided.

When shear strains must be accurately measured and the specimen does not present a symmetry in geometry or loading it is advisable to use a third grating in the Oxy plane subtending a certain angle to the crossed specimen grating. The triaxial grating, with three consecutive superpositions of the *RG*, accurately yields the normal components of displacements u, v and w in the principal directions of the triaxial specimen grating. The normal strains in the u, v and w directions were derived from the relations (2.22), (2.25) and a similar relation for the displacement in the w-direction. The components of strain ε_x, ε_y and γ_{xy} can be derived from the normal components ε_u, ε_v and ε_w by using the well-known strain-rosette relationships. It is advantageous to orient the third grating at 45° to the principal directions of the crossed gratings. In this case the relations expressing the ε_x, ε_y and γ_{xy} components of strain are simplified and coincide with those of the rectangular strain-rosettes,[5] as follows.

[5] Meier, J. H., Strain rosettes, *Handbook of Experimental Stress Analysis*, Hetényi, M. (Ed.), Wiley, New York, 1950.

$$\left.\begin{array}{l} \varepsilon_x = \varepsilon_u, \\ \varepsilon_y = \varepsilon_v, \\ \gamma_{xy} = 2\varepsilon_w - (\varepsilon_u + \varepsilon_v). \end{array}\right\} \tag{2.28}$$

Relations (2.22) and (2.25) show that the intersection points of moiré fringes by parallel lines to the principal axes of the specimen

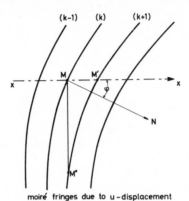

moiré fringes due to u–displacement

FIG. 2.3. Geometry of a local region of moiré fringes related to the principal direction of the reference grating.

do not change position if the reference grating is angularly displaced by a small angle, in spite of the substantial modification of the moiré pattern.

If φ is the angle between the normal to the moiré fringes formed by a deformed SG due to an ε_x strain and the principal direction of RG (Fig. 2.3), then

$$\tan \varphi = \left(\frac{\partial u}{\partial y}\right) : \left(\frac{\partial u}{\partial x}\right).$$

Let the reference grating be angularly displaced by a small angle β. The new slope of moiré fringes, which is no more normal to

x-axis, is given by

$$\tan \varphi' = \left(\frac{\partial U}{\partial y}\right) : \left(\frac{\partial U}{\partial x}\right).$$

Since

$$\frac{\partial U}{\partial y} = \frac{\partial u}{\partial y} + \beta,$$

then

$$\tan \varphi' = \tan \varphi + \beta/\varepsilon_x$$

and

$$\Delta (\tan \varphi) = \frac{\beta}{\varepsilon_x}. \tag{2.29}$$

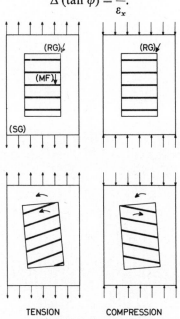

TENSION   COMPRESSION

FIG. 2.4a. Schematic definition of the sign of strain component by a relative angular displacement of the reference grating.

Moiré fringes formed by two gratings $x = pk$ and an applied strain $\varepsilon_x = \partial u/\partial x$ in SG either follow the direction of the angular

FIG. 2.4b, c. Moiré patterns formed by two gratings of different pitch angularly displaced in the same direction and showing the sign of the strain component ((b) specimen grating compressed, (c) specimen grating extended).

displacement of the reference grating if the strain is tensile (positive), or are angularly displaced in an opposite direction to that of *RG* if the strain is compressive (negative) (Fig. 2.4). This rule is useful to disclose the sign of individual strains in a strain field.

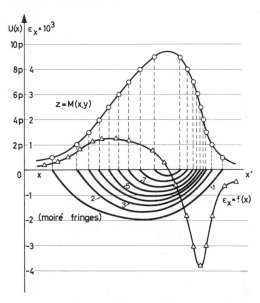

Fig. 2.5. Schematic diagram of the procedure of the graphical derivation of the displacement curve obtained along a traverse of the moiré pattern.

From the definition of moiré fringes, which are the loci of equal relative displacement in the principal directions of the gratings, it can be deduced that moiré fringes are the contour maps of the function $z = M(x,y)$ on the Oxy plane. This function is derived from the functions expressing the families of lines in the two gratings $[R(x,y) = k$ and $S(x,y) = l]$ by applying the indicial relation (1.1). The planes of intersection of the three-dimensional surface $z = M(x,y)$ are given by equation $z = mp$. The surface $z = M(x,y)$ is successively cut by planes normal to the base-plane

(Oxy) and parallel to the principal and secondary directions of the reference grating (Fig. 2.5). If these sections are folded on the base plane, they yield the curves of the variation of displacement along the two principal axes. This displacement versus distance curve represents a u-displacement curve if the gratings have their principal direction parallel to x-axis. Similarly, for the case where the moiré surface $z = M(x,y)$ is formed by gratings with their principal direction parallel to y-axis the sections of the surface $z = M(x,y)$ folded on the base plane yield the curves of variation of the displacement v along the x- and y-axes respectively.

(a)

FIG. 2.6a. Moiré pattern of the transversal displacement field in a perforated strip under tension.

A graphical differentiation of these four curves yields the four components of strains ε_x, γ_{yy}, ε_y, γ_{xx} from which the normal and shear Cartesian strain components are evaluated. The shear strain component γ_{xy} is found by the algebraic addition of the terms γ_{yy} and γ_{xx} derived from the graphical differentiation of the u- and v-curves along the y- and x-axes.

In order to determine the components of strain in a two-dimensional field the displacement gradients in x- and y-directions are required. While the positions of moiré fringes of integral order might be accurately determined, the distance between moiré fringes may be large, thus hindering the accurate evaluation of the local gradients along x- and y-axes. Therefore, accurate measurement of strains necessitates moiré families with closely spaced moiré fringes, which must cross the sections along which the strain components must be evaluated at large angles approaching 90°. These requirements can be achieved with reference and specimen gratings having the same pitch if the measured strains are large.

Figure 2.6a shows the moiré pattern of the transversal displacement field in a perforated strip under tension. The reference grating and the specimen grating before deformation were identical. Since the deformation of the specimen at this loading step was large, the moiré pattern formed by the equal pitch method was a dense pattern allowing the accurate evaluation of displacements.

Figure 2.6b shows a microdensitogram along line *AB*. The

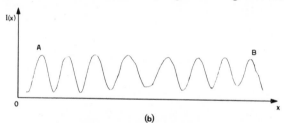

(b)

FIG. 2.6b. Microdensitogram along a longitudinal
section *AB* of the specimen.

maxima and minima of the density distribution curve correspond to the middle points of the white and dark moiré fringes. The accurate definition of the extreme dark and white points of each fringe allows the accurate tracing of the displacement versus distance curve. Since the moiré fringes are contour curves, they represent the variation of the $u(x,y)$ or $v(x,y)$ functions in the same manner as the contours of a topographic map represent the variations of altitude in different points of the soil. The difference in displacement between two consecutive moiré fringes is equal to the pitch of the gratings. Therefore, the moiré pattern corresponds to a continuous vernier system which measures displacements over the whole surface area of the specimen.

For tracing the displacement versus distance curve an arbitrary scale is chosen for the variation of displacement along the section considered, which represents values of displacement expressed in units of pitch. When tracing the $u(x,y)$ or $v(x,y)$ displacement curve versus distance along the section it is convenient to have the

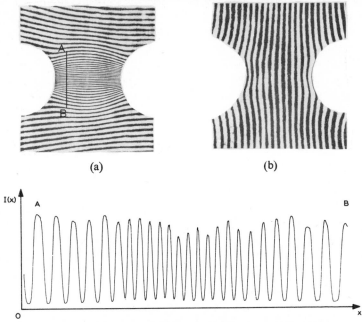

FIG. 2.7. Moiré patterns of the u- and v-displacement fields for a symmetrically grooved aluminium specimen plastically deformed by a longitudinal tension.

abscissas corresponding to distances along the section, while the ordinates represent the variation of displacement.

A graphical differentiation of the u-displacement curve along the section yields the $\partial u/\partial x$ component, while a graphical differentiation of the v-displacement curve along the section yields the $\partial v/\partial x$ component. For the complete evaluation of the strain field a coarse screen of equidistant lines is traced over the specimen and the displacements as well as their derivatives curves versus distances along both families of lines of the screen are graphically evaluated point by point. At each knot of the screen all four components of $\partial u/\partial x$, $\partial u/\partial y$, $\partial v/\partial x$, $\partial v/\partial y$ of the strain field are separately evaluated. From these values the components of strain may be readily calculated

Figure 2.7 (a, b) shows the moiré patterns of u- and v-displace-

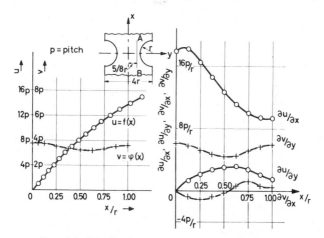

FIG. 2.8. Distribution of the u and v displacements as well as of their derivatives $\dfrac{\partial u}{\partial x}$, $\dfrac{\partial u}{\partial y}$ and $\dfrac{\partial v}{\partial x}$, $\dfrac{\partial v}{\partial y}$ along section AB of a symmetrically grooved aluminium specimen plastically deformed by a longitudinal tension (Fig. 2.7).

ment fields for a symmetrically grooved aluminium specimen subjected to longitudinal tension.

Figure 2.8 shows the $u = f(x)$ and $v = \varphi(x)$ curves along the section AB derived from the moiré patterns of Fig. 2.7. In the same figure the $\partial u/\partial x$, $\partial u/\partial y$ and $\partial v/\partial x$, $\partial v/\partial y$ curves are traced by graphical differentiation. For accurate tracing of the derivative curves it is essential to have the displacement curves traced in great detail. Thence the necessity of having as many moiré fringes as possible to define the displacement curve.

From the above discussion it is clear that the equal pitch moiré method is convenient and yields reliable results in the measurement of large plastic deformations. One can say that the realm of moiré method starts where the realms of the other classical methods of strain analysis, such as photoelasticity, strain gauges, etc., end. It will be shown later on how the domain of moiré method can be extended to measure either small elastic deformations, or, directly, elastic stresses.

In order to increase the sensitivity of the method, gratings of higher frequency must be used. The ideal case may be the use of diffraction gratings, which have a much higher frequency than the ordinary amplitude gratings used in strain analysis. The use of diffraction gratings in moiré analysis encounters many obstacles and difficulties. A main difficulty is in the transfer of the grating configuration from the master to the specimen. For diffraction gratings of a frequency of several thousand lines per inch it is impossible to obtain a good photographic reproduction over an extended surface area of a specimen, even in the simplest case of contact reproduction. This difficulty is followed by the impossibility of obtaining moiré fringes of useful contrast with high frequency diffraction gratings in the simple conditions of illumination and observation, which are normally used in experiments of strain analysis. Moreover, the cost of such gratings is prohibitive for their everyday use in various experiments.

Finally, the moiré patterns formed by diffraction gratings are due to diffraction and interference phenomena obeying complicated laws and any interference based on the obstruction of light alone may be considered as invalid.

A limit frequency for a good photographic reproduction of a master by contact printing on the surface of the specimen is the frequency of 1000 lines per inch (40 lines per millimetre). This limit must be reduced to half the frequency for projection reproduction if uniform illumination conditions are attained and an accurate optical bench for the exact parallelism of the master, the objective, and the specimen planes is at the disposition of the experimenter.

Alternatively, an increase of the linear dimensions of the specimen results in a proportional increase of the sensitivity of the method and circumvents the necessity of increasing the frequency of the gratings since moiré fringes are contours of equal displacement and not equal strain. This principle is in general use in our laboratory, where large specimens are tested with satisfactory sensitivity.

It was previously shown from relations (2.22) and (2.25) that an eventual small angular misalignment does not influence the values of the normal components of strain ε_x and ε_y, although it may change considerably the shape of moiré fringes. On the contrary, the shear components γ_{xx} and γ_{yy} are directly proportional to any eventual angular misalignment of the gratings.

If the angle of misalignment is significant and the displacement components due to misalignment are U_r and V_r respectively, the components of strain are given by relations

$$\varepsilon_x = \frac{\partial U}{\partial x} - R_x,$$

$$\varepsilon_y = \frac{\partial V}{\partial y} - R_y,$$

and
$$\gamma_{xy} = \frac{\partial U}{\partial y} + \frac{\partial V}{\partial x} - (R_{xy} + R_{yx}), \qquad (2.30)$$

where the constants R_x, R_y, R_{xy} and R_{yx} are given by the derivatives

$$\frac{\partial U_r}{\partial x} = R_x, \quad \frac{\partial V_r}{\partial y} = R_y, \quad \frac{\partial U_r}{\partial y} = R_{xy} \quad \text{and} \quad \frac{\partial V_r}{\partial x} = R_{yx}. \qquad (2.31)$$

Again in this case, while the constants R_x and R_y are small and may be neglected, the constants R_{xy} and R_{yx} are large and may introduce substantial errors to the values of shear components.

If the angles of misalignment of the reference grating in the two different directions are equal in magnitude and sense of orientation then R_{xy} and R_{yx} are equal in magnitude and, since the corresponding displacements lie in adjacent u_r and v_r quadrants in the co-ordinate system, their derivatives have opposite signs. Thus, $R_{xy} = -R_{yx}$ and $(R_{xy} + R_{yx}) = 0.$[6] It may then be concluded that the

[6] Diruy, M., *DOCAÉRO*, No. 55, 1 (March 1959).

experimental error in calculating the shear component of strain γ_{xy} by relation (2.30) vanishes not only when angular misalignments become zero, but also when these misalignments in the two orthogonal directions are finite and of equal angle.

Since the components R_{xy} and R_{yx} are constant all over the strain field they may be evaluated at a point such as a free boundary or a free corner of the field, where the shear component of strain is known to be zero. A correction of calculated shear strains all over the field is possible from the correction deduced at one point of the field.

2.3. Measurement of large strains

The basic principle according to which moiré fringes are contour lines of the u- and v-displacements remains rigorously correct for every magnitude of deformation. However, it must be stated that moiré fringes are contour lines of displacement for the deformed specimen. In other words, for a fringe of order n corresponding to a displacement $\Delta u = np$ every point of the fringe is a point M' of the deformed specimen displaced from its initial position by a displacement along the principal direction of the grating equal to $\Delta u = np$.

Indeed, the Eulerian definition of strain referred to the deformed geometry of the body is given by

$$\varepsilon_i^E = \frac{p_i}{f_i}, \qquad (2.32)$$

while the Lagrangean definition of strain referred to the initial geometry of the body is given by

$$\varepsilon_i^L = \frac{p_i}{f_i - p_i}, \qquad (2.33)$$

where p_i and f_i express the corresponding pitches and interfringe spacings of moiré fringes formed by two crossed gratings and i may be taken either as x or y.

The displacement versus distance $(u = f(x))$ curve (UU') traced along a traverse on the x-direction from a moiré pattern must be adjusted to yield the rectified curve $(U_r U_r')$ corresponding to the initial conditions of the specimens (Fig. 2.9). These corrections must be applied in the same manner for the evaluation of partial derivatives $\partial u/\partial x$, $\partial u/\partial y$ and $\partial v/\partial y$, $\partial v/\partial x$.

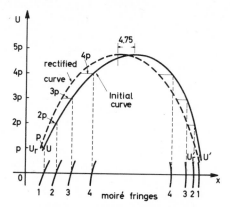

FIG. 2.9. Initial and rectified displacement versus distance curves traced from the corresponding moiré pattern. Rectified curve is adjusted to correspond to the initial conditions of the specimen.

From the rectified curves of displacement versus distance the real values of the derivatives can be derived. These values must be introduced into the general relationship referred to a Lagrangean system of coordinates (Fig. 2.10)

$$
\left.
\begin{aligned}
\varepsilon_x^L &= \left[\left(1+\frac{\partial u}{\partial x}\right)^2+\left(\frac{\partial v}{\partial x}\right)^2\right]^{\frac{1}{2}}-1, \\
\varepsilon_y^L &= \left[\left(1+\frac{\partial v}{\partial y}\right)^2+\left(\frac{\partial u}{\partial y}\right)^2\right]^{\frac{1}{2}}-1, \\
\gamma_{xy}^L &= \arc\sin \frac{\dfrac{\partial u}{\partial y}+\dfrac{\partial v}{\partial x}+\dfrac{\partial u}{\partial x}\cdot\dfrac{\partial u}{\partial y}+\dfrac{\partial v}{\partial x}\cdot\dfrac{\partial v}{\partial y}}{\left(1+\varepsilon_x^L\right)\cdot\left(1+\varepsilon_y^L\right)}.
\end{aligned}
\right\}
\qquad (2.34)
$$

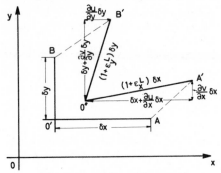

FIG. 2.10. Evaluation of large strain components in a Lagrangean system of coordinates in terms of displacements u and v in the Oxy plane.

An approximation of eqns. (2.34), in which infinitesimals of higher than the first order are neglected, is given by the following relations:

$$
\left.
\begin{aligned}
\varepsilon_x^L &= \left[1+2\frac{\partial u}{\partial x}\right]^{\frac{1}{2}}-1, \\
\varepsilon_y^L &= \left[1+2\frac{\partial v}{\partial y}\right]^{\frac{1}{2}}-1, \\
\gamma_{xy}^L &= \arcsin \frac{\dfrac{\partial u}{\partial y}+\dfrac{\partial v}{\partial x}}{\left(1+\varepsilon_x^L\right)\left(1+\varepsilon_y^L\right)}.
\end{aligned}
\right\}
\qquad (2.35)
$$

The rectification of the displacement versus distance curves traced from the moiré patterns and the use of relations (2.34) for large strains allow a complete and rapid correction of values of strains in cases where the deformations are sufficiently large and the second partial derivatives of u and v with respect to x and y cannot be considered as negligible.

2.4. The linear differential moiré method

The equal pitch moiré method is only suitable for large strain fields. Dense moiré patterns are formed by this method with sharp fringes allowing an accurate evaluation of the strain field. In the case where small elastic strains are to be evaluated the equal pitch method becomes invalid as it yields patterns with sparse and irregular fringes, which cover large areas of the strain field and they are unsuitable for a moiré strain analysis. In this case the differential or mismatch method is adequate. Diruy[7] was the first to mention the possibility of using the differential moiré method for elastic strain analysis. An extensive use of the differential method in various problems of elasticity, plasticity and visco-elasticity, as well as a full development of the theory of the method, is given by Theocaris and his co-workers[8]. More recently Durelli and Sciammarella[9] used the differential method to solve a problem of elastoplastic stress and strain distribution around a hole, while Sciammarella and Chiang[10] applied it for the analysis of the classical Hertz contact problem.

The method is based on the use of a series of reference gratings of a different pitch than that of the original pitch of the specimen grating. Reduced-pitch line reference gratings are appropriate to measure extensional strain fields while expanded-pitch gratings are suitable for contractional fields. By using a reference grating (*RG*) of slightly different pitch than that of the specimen grating (*SG*) a fictitious initial deformation is introduced which results in dense and well-defined moiré pattern although the pattern due to

[7] Diruy, M., *DOCAÉRO*, No. 55, 1 (March 1959).

[8] Theocaris, P. S., *Proc. Nat. Acad. Athens* **36**, 238 (1961). See also Theocaris, P. S., *Proc. Am. Soc. Test. Mater.* **61**, 838 (1961). Theocaris, P. S., and Koroneos, E., *Phil. Mag.* **8** (95) 1871 (1963); *Proc. Am. Soc. Test. Mater.* **64**, 747 (1964). Theocaris, P. S., and Marketos, E., *Proc. Int. Conf. Fracture, Sendai, Japan* **2**, 178 (1965), *Acta Mech.* **4** (1967). Theocaris, P. S., *Proc. Soc. Exp. Stress Anal.* **21** (2) 289 (1964), and Theocaris, P. S., and Hadjijoseph, C., *Kolloid-Z.* **202** (2) 133 (1965).

[9] Durelli, A. J., and Sciammarella, C., *Trans. ASME, J. Appl. Mech.* **30** (1) 115 (1963).

[10] Sciammarella, C., and Chiang, Fu-Pen, *Exp. Mech.* **4** (2) 313 (1964).

the actual deformation of the specimen contains only few and ill-defined fringes. Since the composite moiré pattern contains close-spaced and sharp fringes, the tracing of the displacement versus distance curves becomes accurate and easy. The fictitious displacement introduced by the disparity in pitch between RG and SG is linearly proportional to the coordinates and the resulting strain is constant all over the field. This constant strain may be subtracted from the total values of strains derived from the composite displacement field.

For an initial difference in pitches of RG and SG this disparity is called linear disparity. If the pitch of RG is greater than the pitch of SG the disparity is called compressive disparity or mismatch, since it corresponds to a fictitious initial compressive strain and it is suitable for measuring compressive components of strain. If the pitch of RG is smaller than the pitch of SG the disparity is called tensile disparity. Besides the linear disparity another type of disparity exists due to an angular displacement of RG. This is called angular disparity or mismatch.

While linear disparities increase only the fringe density along principal directions, angular disparities increase the fringe density along secondary directions. Thus, while linear disparities are suitable for measurements of normal components of strain, angular disparities are preferable for measuring γ_{xx} and γ_{yy}, i.e. the angular components yielding the shear strain γ_{xy} of the strain field.

If both linear and angular disparities are contributing to a moiré pattern it is easy to show that for λ-positive (compressive disparity) and an angle of angular displacement ϑ-positive clockwise, a positive slope of moiré fringes is engendered, while for λ-negative (tensile disparity) and ϑ-positive a negative slope of moiré fringes is created (angle φ formed by the principal directions of the specimen grating and the moiré fringes is either acute or obtuse in these cases (Fig. 2.4)).

If the pitch of the reference grating is equal to $p(1 + \lambda)$, where λ obtains positive or negative infinitesimal values, it is suitable to choose λ-negative for extensional strain fields and λ-positive for contractional strain fields. In this manner the fictitious initial deformation is added to the actual deformation of the specimen and the interfringe spacing diminishes with further increase of the external load. Increasing the values of λ causes a decrease of the initial interfringe spacing. In the case where the mode of displacement along one direction of the specimen is not known in advance a free choice of the sign in λ does not influence the accuracy of the method. The result will be the same with the exception that, instead of a diminishing interfringe spacing with a load increase, an increase of the interfringe spacing will be observed. In this case, when the fringe density becomes critical to the accuracy of the evaluation of strain components, the reference grating can be substituted by one having the appropriate pitch.

The transition from the reference grating to the deformed specimen grating may be accomplished by two successive transformations:

(a) an affinity of the relation $p/p' = 1/(1+\lambda_x) \cong (1-\lambda_x)$, if terms in λ_x^2 are considered negligible; and

(b) a displacement u corresponding to the deformation of the body.

The affinity p/p' is equivalent to a fictitious displacement $u_f = -\lambda_x x$ and thus the total displacement U, given by the moiré pattern, is defined as

$$U = u + u_f = u - \lambda_x x. \tag{2.36}$$

Therefore, the moiré fringes of the composite moiré pattern are the contour curves of the function $U(x,y)$. The displacement between two successive moiré fringes is, in this case, equal to $p(1+\lambda_x)$.

The component ε_x of strain is formed by differentiating relation (2.36), i.e.

$$\varepsilon_x = \frac{\partial u}{\partial x} = \frac{\partial U}{\partial x} + \lambda_x. \tag{2.37}$$

Similarly, for the direction at right angles the component of strain is given as

$$\varepsilon_y = \frac{\partial v}{\partial y} = \frac{\partial V}{\partial y} + \lambda_y, \tag{2.38}$$

where λ_x and λ_y are the disparities in pitch of the reference gratings with principal directions parallel to the x- and y-axes respectively.

It was previously shown [relations (2.23) and (2.26)] that the shear components γ_{xx} and γ_{yy} of the total shear strain γ_{xy} are un-affected by the introduction of the initial fictitious displacement fields (linear disparities). These components may be determined directly from the corresponding moiré patterns if care is taken to avoid accidental angular misalignments during the superposition of each reference grating on the specimen surface.

The moiré patterns formed by a crossed specimen grating to which two reference gratings with slightly different pitches are successively superposed yield the displacement curves from which the shear strain components may be evaluated without correction. The normal components may be found by algebraically adding to the total strain components the constant quantities λ_x and λ_y, expressing the disparities in pitch of the two reference gratings. The algebraic addition of the constant components of strain λ_x and λ_y may be executed either graphically or numerically.

The possibility of varying the quantity λ by disposing a series of reference gratings with various values of λ yields a further advantage to the differential method. It is possible to properly choose certain values of λ in a manner to yield moiré patterns with various densities at different areas of the strain field. Indeed, the fringe spacing is varying rapidly with λ. This spacing may, for a certain value of λ, increase in some areas and decrease in other areas. This fact allows a significant increase of the points of measurement, which increase considerably helps the graphical differentiation of the displacement versus distance curves. Figure 2.11 shows the moiré pattern of a circular ring diametrically compressed when a reference grating of $p = 0.002\,(1+\lambda)$ in. is used. Figure 2.11a corresponds to a $\lambda = 0.012$, while Fig. 2.11b shows the moiré pattern for $\lambda = 0.036$ in./in. The traced black lines on the moiré photographs represent the ε_x-isoentatics corresponding to the two values of λ.

The differential moiré method resembles the compensation method of classical photoelasticity because it provides additional fringes obtained by using a series of reference gratings of slightly different pitch. These fringes lie between the fringes formed by the equal pitch method and, hence, divide the length between two fringes into smaller intervals. Thus either the gauge length is reduced or, for the same gauge length, the sensitivity of the method is increased.

For the evaluation of the components of strains by the differential method the total displacement $[U(x,y)$ or $V(x,y)]$ versus distance curves are traced and the total strains due to the fictitious

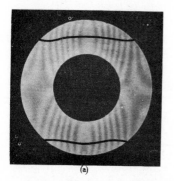
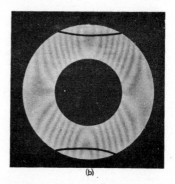

FIG. 2.11. Moiré patterns for a diametrically compressed circular ring formed by the linear differential method: (a) $\lambda = 0.012$ in./in.; (b) $\lambda = 0.036$ in./in.

initial displacement field together with the deformations of the specimen are evaluated by graphical differentiation of the corresponding total displacement curves. For the case of normal components of strains, they are reduced to their real values by algebraically adding the quantities λ_x or λ_y which are evaluated from a record of the fringe pattern with the specimen unloaded. If there is no rotational disparity in pitches of the gratings the shear components γ_{xx} and γ_{yy} are unaffected by the linear disparities and therefore the values of these components derived from the corresponding displacement versus distance curves are the values of these components due to the deformation of the specimen.

There are two possibilities of application of the differential moiré method.

(a) A series of reference gratings of slightly reduced or enlarged pitches from the standard pitch, selected for the specimen grating, must be prepared on glass plates by photographic reduction on a photo-engraver's bench. Crossed gratings for SG are photographically prepared by contact printing from the standard line grating by twice exposing the film along the principal and secondary directions of the grating. The reproduction of the master crossed grating on the surface of the specimen is made by one of the photo-engraving processes described in the appropriate chapter. The moiré pattern is formed by the superposition of each reference grating on the gauge area of the specimen where the crossed grating is printed. The pattern is illuminated and photographed by a camera placed at right angle to the axis of the specimen (contact method).

(b) It is also possible, if the frequency of SG is not high (up to 500 lines per inch), to form the image of the crossed grating on the ground-glass screen of a camera. The camera lens must have a suitable resolving power in order to reproduce an undistorted image of the crossed grating on the ground-glass screen at unit magnification. The image of SG interferes with RG juxtaposed to the ground-glass screen, which, when properly oriented, forms a moiré pattern. If the magnification of the image of SG is equal to unity and the pitch of RG equals the pitch of SG for a highly corrected lens no moiré fringes are formed on the ground-glass screen. If the camera back is slightly approached to or receded from its zero fringe position, moiré fringes appear due to the disparity in pitch between RG and the image of SG, which is, in this case, the varying pitch grating. The values of λ corresponding to each position of the camera back may be readily determined by elementary optics if the infinitesimal displacement of the ground-glass screen is accurately measured with the help of a vernier system. If the movement of the ground-glass screen is beyond the limits of the depth of field of the objective, it will be necessary to slightly displace the objective in order to refocus the grating. This movement corresponds to a change in magnification of SG. The displacement may be measured with high accuracy and the new relative position of the specimen, lens and RG may easily yield the value of λ (image method).

2.5. Direct tracing of isoentatics

Consider a family of moiré fringes corresponding to a moiré pattern of the u-displacements and formed by a reference grating of pitch $p(1+\lambda)$ (Fig. 2.3). If MN is the principal direction of fringe k at point M and φ the angle subtended to the x-axis, from the moiré theory, it is valid for the displacements between two successive moiré fringes (k) and $(k+1)$ that

$$U(M') - U(M) = (MM')\frac{\partial U}{\partial x} = p(1+\lambda),$$

or

$$(MM') = p(1+\lambda) : \left(\frac{\partial U}{\partial x}\right).$$

Similarly,

$$(MM'') = p(1+\lambda) : \left(\frac{\partial U}{\partial y}\right).$$

Therefore,

$$\tan \varphi = \left(\frac{\partial U}{\partial y}\right) \Big/ \left(\frac{\partial U}{\partial x}\right). \tag{2.39}$$

The points where the tangents to moiré fringes are parallel to either of the x, y-axes are given by the relations (Fig. 2.12)

$$\frac{\partial U}{\partial x} = 0 \quad \text{for } x\text{-axis,}$$

$$\frac{\partial U}{\partial y} = 0 \quad \text{for } y\text{-axis.}$$

Moreover, the component ε_x of strain is given by

$$\varepsilon_x = \frac{\partial U}{\partial x} + \lambda. \tag{2.40}$$

The strain ε_x is equal to λ at the points where $\partial U/\partial x = 0$. These are points of contact of the tangent parallel to x-axis and the corresponding moiré fringe.

By varying λ it is possible to trace the lines along which the tangents to the moiré fringes are parallel to the principal axes of the specimen. These lines are the ε_x or ε_y isoentatics. By disposing a series of references gratings with regularly varying values of λ it is possible to trace the isoentatics in a strain field.

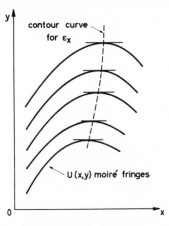

FIG. 2.12. Tracing of $\varepsilon_x = \lambda$ isoentatic curve.

In order to trace the γ_{xy} isoentatics a reference grating of an equal pitch to that of the specimen grating is needed. If RG is angularly displaced by a small angle ϑ, a moiré pattern will be formed with fringes almost normal to x-axis if a moiré pattern of u-displacements is studied and the specimen is unloaded. With the specimen loaded the total displacement U will contain a component u parallel to x-axis, derived from the angular displacement ϑ $(u_r = \vartheta y)$.

Then,
$$\frac{\partial U}{\partial y} = \frac{\partial u}{\partial y} + \vartheta. \tag{2.41}$$

At points where $\partial U/\partial y = 0$ it is valid that $\partial u/\partial y = -\vartheta$. Therefore, it is possible to trace the contour curve $\partial u/\partial y$ by connecting the points of moiré fringes in the loaded pattern formed by an angular displacement ϑ of RG. By varying the angle ϑ a series of $\partial u/\partial y$ contour curves may be traced. Similarly, for a moiré pattern corresponding to v-displacements a series of $\partial v/\partial x$ contour curves may be traced by varying the angle of rotation of RG. From the two families of the $\partial u/\partial y$ and $\partial v/\partial x$ contour curves the shear component of strain may be evaluated all over the strain field.

It is therefore possible by systematically varying either the pitch of the reference grating or the angular displacement to obtain the contour curves of the strain components in the specimen.

The method has been indicated by Dantu[11] and applied to an elasticity problem by Theocaris.[12] This example is concerned with the strain distribution in a circular ring diametrically compressed in a state of plane stress.

The specimen was prepared from a cold-setting pure epoxy polymer sheet 0·25 in. thick. The external diameter of the ring was $D = 42$ mm and the internal diameter $d = 21$ mm. The lateral surfaces of the ring were polished and on one of them a crossed grating of a line frequency of 500 lines per inch was reproduced. The diametral compression was applied in a direction coinciding with the principal direction of one of the gratings when the polymer was at its rubbery state of 135°C. The deformation of the specimen was frozen by cooling the specimen to ambient temperature in its loaded state. A photographic copy of the deformed crossed grating was obtained at unit magnification and this copy replaced the true specimen in all further procedures. The photographic copy of the deformed specimen was placed on the ground glass of an intense white diffuse light source and constituted the specimen grating.

For convenience, the image differential moiré method was used

[11] Dantu, P., Laboratoire Central des Ponts et Chaussées, Publ. No. 57–6 (1957).
[12] Theocaris, P. S., *Materialprüfung*, **10** (5) 155 (1968).

as it was developed by Theocaris.[13] According to this method, the image of the specimen grating was formed at unit magnification on the ground-glass screen of a still camera and the reference grating interfered with the image of the specimen grating forming a moiré pattern. This moiré pattern was photographed by a second still camera. In this manner only one reference grating was used. By slightly increasing or diminishing the magnification of the image of the specimen grating it was possible to establish different values of the term λ of the reference grating. From ele-

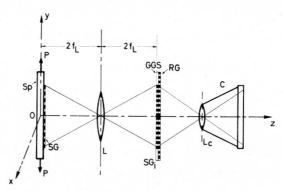

Fig. 2.13. Schematic representation of the set-up used in the image moiré method.

mentary optics of lenses it can easily be shown that the displacements of the ground-glass screen of the first still camera, which were small in order to introduce infinitesimal values for λ, were directly proportional to the magnification sought for the specimen grating. By establishing a unit displacement for the ground-glass screen and the reference grating and increasing this displacement by equal amounts each time, isoentatics of equal increments of strain were obtained. A schematic representation of the set-up used in the experiments is shown in Fig. 2.13.

By angularly displacing the reference grating through 90°, at each displacement step of *RG* it was possible to obtain the isoenta-

[13] Theocaris, P. S., *Exp. Mech.* **4** (8) 223 (1964).

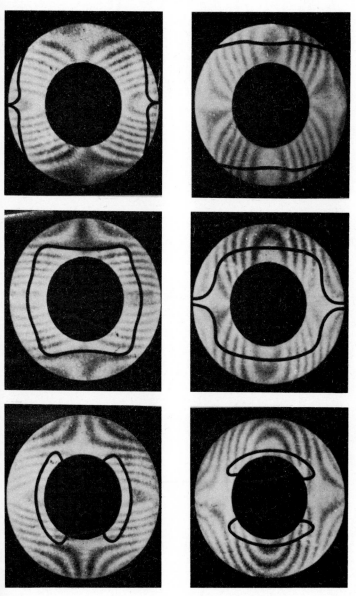

Fig. 2.14. Moiré patterns of the u- and v-displacements for three different values of λ.

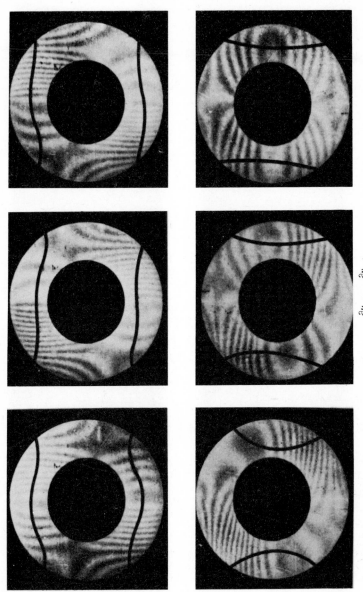

FIG. 2.15. Moiré patterns of both cross derivatives $\frac{\partial u}{\partial y}$ and $\frac{\partial v}{\partial x}$ of displacements for three different angular displacements of the reference grating.

tics of the other normal component of strain without any other change in the set-up. Figure 2.14 shows the moiré patterns of the u- and v-displacements for three different values of λ. In these patterns the corresponding isoentatics were traced by applying this technique.

In order to trace the cross derivatives $\partial u/\partial y$ and $\partial v/\partial x$ of the displacements and evaluate the shear strain component γ_{xy}, the first still camera was adjusted at unit magnification, the moiré

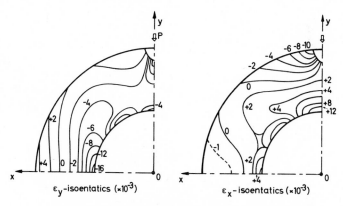

FIG. 2.16. The ε_x and ε_y isoentatics for a diametrically compressed circular ring derived from a series of moiré patterns.

pattern corresponding to the deformation of the ring in the case where the rulings of RG and SG were parallel. By angularly displacing the reference grating by a small angle, a series of different moiré patterns were formed after each additional increment of angular displacement which yielded the values of the one cross derivative (say $\partial u/\partial y$). By displacing RG through $90°$ and repeating the procedure, another series of moiré patterns were formed which yielded the other cross derivative ($\partial v/\partial x$). An algebraic addition of the values of the two cross derivatives yielded the shear isoentatics. Figure 2.15 shows the moiré patterns of both cross derivatives of displacements for three different angular displacements.

Figures 2.16 and 2.17 show the ε_x and ε_y isoentatics and γ_{xy} isoentatics as these were derived from a series of moiré patterns. Comparison of the experimentally obtained curves with either the theoretical values or the experimental values derived from photo-elasticity showed a remarkable coincidence.

While the tracing of a single isoentatic by this method may not

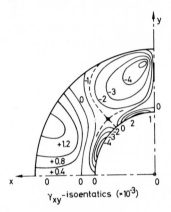

FIG. 2.17. The γ_{xy} isoentatics for a dia-metrically compressed circular ring derived from the cross deriva-tives of the displacement components.

be very accurate since the shape of the moiré fringes does not change abruptly and the tangents to the fringes cannot be accur-ately determined, the tracing of a whole family of isoentatics of a strain component covering the whole field is much more accurate since the tracing of the isoentatic is helped by the tracing of its neighbours especially in regions where the direction of the particu-lar isoentatic is unaltered over the area of the specimen.

Moreover, the accuracy of the method may be further improved by increasing the line frequency of the gratings. An alternative means of increasing the accuracy and the sensitivity of the method in the case where the possibilities of increasing the line frequency of the gratings are exhausted, is to increase the size of the specimen since moiré fringes are contours of equal displacement and not of equal strain and therefore an increase of the linear dimensions of the specimen results in a proportional increase of the sensitivity of the method.

2.6. The angular disparity moiré method

Besides the linear disparity method described in detail previously, the angular disparity or mismatch method can be used for the determination of the direct and shear strain components of a field.

The formation of fringes by two identical gratings having a relative angular displacement was described in Chapter 1. Dantu,[14] Morse, Durelli and Sciammarella,[15] and Low and Bray[16] have described the formation of fringes due to an angular displacement and have derived expressions interrelating the main geometric parameters, i.e. relative angular displacement of gratings, fringe angular displacement, interfringe spacing, Cartesian components of interfringe spacing, etc. Parks and Durelli[17] and Chiang[18] have indicated that from measurements of fringe location the u- and v-components of displacements along the x- and y-axes are determined, from which the partial derivatives $\partial u/\partial x$, $\partial u/\partial y$, $\partial v/\partial y$ and $\partial v/\partial x$ can be evaluated. The direct and shear strain components can be evaluated by employing, according to the definition of strain (Eulerian or Lagrangean), the various strain-displacement relationships.

The more direct application of angular displacement of fringes to the determination of strain was the method developed by

[14] Dantu, P., Lab. Cent. Ponts Chauss., Publ. No. 57–6 (1957).
[15] Morse, S., Durelli, A. J., and Sciammarella, C. A., *Proc. ASCE, J. Engng. Mech. Div.* **86** (7) 105 (1960).
[16] Low, I. A. B., and Bray, J. W., *The Engineer* **213** (5) 566 (1962).
[17] Parks, V. J., and Durelli, A. J., *Exp. Mech.* **4** (2) 37 (1964).
[18] Chiang, F. P., *Proc. ASCE, J. Engng. Mech. Div.* **91** (1) 137 (1965).

Tanaka and Nakashima,[19] recently revised and illustrated by Vafiadakis and Lamble[20].

When the reference grating is superposed on the specimen grating and is angularly displaced through an angle ϑ, a deformation will cause a change in the specimen grating pitch, which will be represented in the moiré pattern by an angular displacement of the fringes from an initial angle φ_1 to a final angle φ_2, where φ is the angle between the fringes and the coordinate axis.

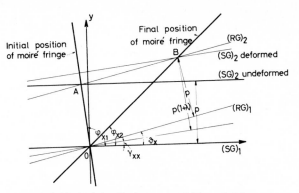

FIG. 2.18. Angular displacement of fringe OA due to a linear and an angular displacement of the gratings.

By rearranging eqn. (2.9) it can be deduced that a strain normal to the deformed specimen grating is given by

$$\varepsilon_y = \frac{p(1+\lambda)}{p} - 1 = \frac{\sin \varphi_2}{\sin (\varphi_2 - \vartheta_x)} - 1. \tag{2.42}$$

By choosing small values of ϑ and measuring the corresponding angular displacement of the fringes, strain can be determined from the above expression without necessitating the measurement of interfringe spacing.

In the more general case where the applied deformation angularly

[19] Tanaka, K., and Nakashima, M., *Proc. Fifth Japan Cong. Test. Mater.* 82 (1962).

[20] Vafiadakis, A. P., and Lamble, J. H., *J. Strain Anal.* **2** (2) 99 (1967).

displaces the rulings of SG through an angle γ_{xx} (Fig. 2.18), it can be shown that strain normal to the deformed grating is given by

$$\varepsilon_y = \frac{\sin(\varphi_2 - \gamma_{xx})}{\sin(\varphi_2 - \vartheta_x)} - 1. \tag{2.43}$$

Choosing two values of ϑ (say $\pm \vartheta = $ constant) and measuring the corresponding fringe inclinations, ε_y and γ_{xx} can be determined by solving the resulting system of equations.

Similarly, by repeating the procedure for a second grating at right angles to the first one, ε_x and γ_{yy} can be determined. Therefore, from four values of fringe angular displacement the complete state of strain at a point can be established, i.e. ε_y, ε_x and γ_{xy}, where $\gamma_{xy} = (\gamma_{xx} + \gamma_{yy})$.

For large angular displacements γ_{xx} and γ_{yy}, strains ε_x and ε_y can be referred to the actual direct strains.

From the relationships (2.42) and (2.43) it can be shown that:

(a) The magnitude of strain that can be measured is independent of the pitch of gratings and solely dependent on the initial relative angular displacement of the gratings. For small values of ϑ (up to 2°) strains of elastic magnitude can be measured.

(b) The sign of strain (i.e., whether tensile or compressive) is directly determined from the expressions of strain and does not necessitate a knowledge of boundary conditions or the sign of strain at one point in the field.

(c) The value of ϑ must be established within few minutes of arc in order to minimize errors in the values of deformation.

(d) By using protractors, values of φ can be measured to within half a degree. For a more accurate determination of deformation better techniques in establishing the fringe inclination must be developed.

(e) The gauge length can be considered equal to the line pitch. In practice about 10 lines are required to clearly define a fringe at a point.

The advantage the angular disparity technique has over the linear disparity technique is that it yields fringes along the secondary grating directions, from which the angular displacement of the specimen grating rulings can be easily determined.

The technique of direct determination of strain from fringe angular displacement does not necessitate the tedious measurements of fringe centre-line and interfringe spacing, the essential differentiation of the displacement components and the calculations of strain from the strain-displacement relationship. From a nomogram of strain versus fringe inclinations for various values of angular displacement of the reference grating and two particular values of ϑ (say $\vartheta = \pm\ 2°$), both ε_x and γ_{yy} or ε_y and γ_{xx} can be read off from the measured values of fringe inclination. It has been illustrated[21] that as the gauge length is not a function of disparity or deformation, as in the linear disparity method, it can be considered as infinitesimal and therefore, since the fringes run in the directions that strains are to be measured, the technique can easily determine non-homogeneities in a strain field and accurately determine the location of strain discontinuous fronts.

2.7. The grid-analyser moiré method

Another approach to obtain the complete strain solution, besides the two gratings approach for symmetric strain fields and the triaxial plane grating approach for the most general case of nonsymmetric strain fields, is the so-called grid-analyser moiré method. This method has been developed and exemplified by Post.[22] According to this technique orthogonally crossed reference and specimen gratings are superposed to yield a complex moiré pattern which contains all the necessary information for a complete analysis of the strain components in a two-dimensional elastic field. The resulting intertwining of the two families of moiré fringes formed by the superposition of the two gratings, which results in a disguising of the identity of each family, is prevented by the use of initial patterns due to linear and angular disparities. The initial patterns

21 Vafiadakis, A. P., and Lamble, J. H., *J. Strain Anal.* **2** (2) 99 (1967).
22 Post, D., *Exp. Mech.* **5** (11) 368 (1965). See also Parks, V. J., *Exp. Mech.* **6** (5) 287 (1966).

have the advantage of reducing any uncertainty in assignment of moiré fringes throughout the field and they provide numerous data points in any local region for reliable evaluation of the fringe gradients.

While the use of crossed specimen and reference gratings in its simplest form may lead to a confusion of the fringe parameters, it is possible to circumvent this deficiency by using an initial pattern formed either by linear disparity in pitch or by angular disparity or by both disparities (Fig. 2.19). When initial patterns are intro-

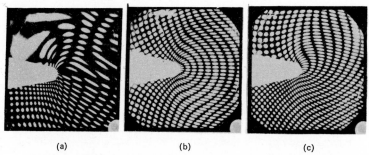

(a)	(b)	(c)

FIG. 2.19. Moiré patterns of crossed reference and specimen gratings in the same strain field created in a deeply notched tensile bar and corresponding to (a) a small angular disparity between gratings, (b) a small disparity in pitch, and (c) a combined angular and linear disparity in pitch. (*Courtesy of Dr. D. Post.*)

duced in the specimen, the displacements due to a subsequently applied strain field are added to the initial displacements and, if the fictitious displacements of the initial pattern are at least a few times greater than the displacements introduced by the strain field, the final pattern will exhibit the general features of the initial pattern, i.e. the final pattern will appear with closely spaced and inclined fringes.

Consider a system of X and Y Cartesian coordinates related to the undeformed specimen. The angular displacement ω of the specimen grating due to a rigid-body rotation of the deformed specimen defines a new coordinate system x', y' related to the first by the angle ω. When the specimen is deformed the coordinate system x', y' as well as the specimen grating will be deformed to a new coordinate system x, y and a grating shown in Fig. 2.20.

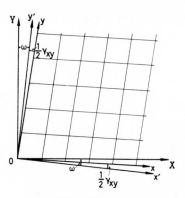

F$_{IG}$. 2.20. Geometry of a local region of the deformed specimen grating.

Angle γ represents the shear strain and the angle ω the rigid-body rotation for this region of the specimen. Angle ω is

$$\sin \omega = \frac{1}{2}\left(\frac{\partial v}{\partial X} - \frac{\partial u}{\partial Y}\right) \qquad (2.44)$$

while the shear strain γ_{xy} is given by

$$\gamma_{xy} = \left(\frac{\partial u}{\partial y} + \frac{\partial v}{\partial x}\right). \qquad (2.44a)$$

Since the shear strain γ_{xy} is small everywhere the x', y' coordinate system is, in practice, indistinguishable from the x, y system and may be considered as coinciding with it. Similarly, the gradients of moiré fringes in any direction of both systems may be taken as identical.

Let p be the pitch of both X and Y specimen gratings prior deformation and $p(1 + \lambda)$ the pitches of the crossed reference grating in both X- and Y-directions.

Superpose the reference grating upon the specimen with the

rulings parallel to the rulings of the specimen grating and then angularly displace the reference grating to form the desired gradient of initial moiré fringes. Define this angular displacement relative to the Cartesian x'- and y'-axes by an angle ϑ-positive for a clockwise angular displacement of the reference grating. The condition of uniformly spaced rulings in an element of the body can be applied

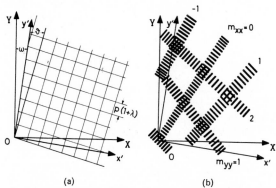

(a)   (b)

Fig. 2.21. (a) Geometry of a local region of the crossed reference grating presenting a linear λ and an angular disparity γ when superposed to a deformed and angularly displaced crossed specimen grating. (b) The two families of moiré fringes formed by the superposition of the deformed specimen grating and the reference grating.

without loss of generality. The superposition of the reference and specimen gratings forms two families of straight and uniformly spaced fringes in the neighbourhood of the element (Fig. 2.21). A strain gradient in the field results in an irregular spacing of moiré fringes.

The moiré fringe orders m_{xx} and m_{yy} are given by the relations (see eqn. 1.1)

$$m_{xx} = k_{xx} - l_{xx}, \qquad (2.45)$$

$$m_{yy} = k_{yy} - l_{yy}. \qquad (2.46)$$

In the following analysis the x-family of rulings in both gratings will be considered. The relations derived from this family of rulings are also valid for the second family of rulings at right angles by interchanging the coordinate x by y. At any point on the x'-axis

$$m_{xx}f_{xx} = l_{xx}p_{sxx} = k_{xx}p_{rxx}. \tag{2.47}$$

From eqns. (2.45) and (2.47)

$$f_{xx} = \frac{p_{rxx}p_{sxx}}{p_{sxx} - p_{rxx}}, \tag{2.48}$$

where f_{xx} is the interfringe spacing of the family of moiré fringes formed by gratings with the x-axis as principal direction and p_{rxx} and p_{sxx} are the pitches of the reference and specimen gratings. The reference and the specimen pitches after deformation are given by relations

$$p_{rxx} = p\frac{(1+\lambda)}{\cos \vartheta} \tag{2.49}$$

and

$$p_{sxx} = p.(1+\varepsilon_x), \tag{2.50}$$

where ε_x is the component of strain along the x-axis.

Introducing eqns. (2.49) and (2.50) into eqn. (2.48) the interfringe spacing f_{xx} is given by

$$f_{xx} = \frac{-p(1+\varepsilon_x)}{1 - \left(\dfrac{1+\varepsilon_x}{1+\lambda}\right)\cos \vartheta}. \tag{2.51}$$

Within the limitations of small-strain theory the $U(x,y)$ total displacement at any point in the x-direction due to a linear and angular disparity in pitch of the reference grating, as well as due to a rigid body rotation of the specimen, is

$$U = pm_{xx}, \tag{2.52}$$

and

$$\frac{\partial U}{\partial x'} = \frac{p}{f_{xx}}. \tag{2.53}$$

By combining eqns. (2.51) and (2.53) and neglecting terms with

ε_x^2 when subtracted to unit, it is deduced that

$$\varepsilon_x = 1 - \frac{\cos \vartheta}{(1+\lambda)} + \frac{\partial U}{\partial x'}.$$

By eqn. (2.52) it is valid that

$$\varepsilon_x = 1 - \frac{\cos \vartheta}{(1+\lambda)} + p\frac{\partial m_{xx}}{\partial x'}. \tag{2.54}$$

Similarly,

$$\varepsilon_y = 1 - \frac{\cos \vartheta}{(1+\lambda)} + p\frac{\partial m_{yy}}{\partial y'}. \tag{2.55}$$

Referring to the specimen grating with principal direction the x-axis its pitch in the y-direction is zero, since the rulings do not cross the y-axis. However, in the y'-direction it is valid that

$$p_{sxy} = \frac{2p(1+\varepsilon_x)}{\gamma_{xy}} \tag{2.56}$$

and

$$p_{rxy} = \frac{p(1+\lambda)}{\sin \vartheta}. \tag{2.57}$$

Let positive shear γ_{xy} be defined as the angular displacement which results in an acute angle between the x- and y-axes. A positive angular disparity ϑ of RG corresponds to a clockwise rotation of the reference grating with respect to x'- and y'-axes. With these definitions, positive shear strain on the specimen grating diminishes the interfringe spacing in the y'-direction whereas positive angular disparity has the opposite effect. Accordingly,

$$f_{xy} = \frac{p_{sxy}p_{rxy}}{p_{rxy} - p_{sxy}}. \tag{2.58}$$

Introducing relations (2.56) and (2.57) into eqn. (2.58) the interfringe spacing f_{xy} is given by

$$f_{xy} = \frac{p(1+\varepsilon_x)}{\frac{1}{2}\gamma_{xy} - \left(\frac{1+\varepsilon_x}{1+\lambda}\right)\sin \vartheta} \tag{2.59}$$

For small strains

$$\frac{\partial U}{\partial y'} = \frac{p}{f_{xy}}. \tag{2.60}$$

Combining eqns. (2.59) and (2.60) yields that

$$\frac{\partial U}{\partial y'} = \frac{\gamma_{xy}}{2(1+\varepsilon_x)} - \frac{\sin \vartheta}{(1+\lambda)}, \tag{2.61}$$

and from eqn. (2.52)

$$p\frac{\partial m_{xx}}{\partial y'} = \frac{\gamma_{xy}}{2(1+\varepsilon_x)} - \frac{\sin \vartheta}{(1+\lambda)}. \tag{2.62}$$

Referring to gratings with principal direction the y-axis,

$$p_{syx} = \frac{2p(1+\varepsilon_y)}{\gamma_{xy}} \tag{2.63}$$

and

$$p_{ryx} = \frac{p(1+\lambda)}{\sin \vartheta}. \tag{2.64}$$

Since the interfringe spacing f_{yx} in the x'-direction is decreasing with positive angular displacement for both γ_{xy} and ϑ, it is valid that

$$f_{yx} = \frac{p_{syx}p_{ryx}}{p_{ryx}+p_{syx}}$$

since

$$V = pm_{yy};$$

also

$$\frac{\partial V}{\partial x'} = p\frac{\partial m_{yy}}{\partial x'} = \frac{\gamma_{xy}}{2(1+\varepsilon_y)} + \frac{\sin \vartheta}{(1+\lambda)}. \tag{2.65}$$

Adding or subtracting eqns. (2.61) and (2.64) and neglecting ε_x and ε_y as infinitesimals compared to unit,

$$\frac{\partial U}{\partial y'} + \frac{\partial V}{\partial x'} = p\left(\frac{\partial m_{yy}}{\partial x'} + \frac{\partial m_{xx}}{\partial y'}\right),$$

$$\frac{\partial V}{\partial x'} - \frac{\partial U}{\partial y'} = \frac{2 \sin \vartheta}{(1+\lambda)}.$$

Accordingly,

$$\gamma_{xy} = p\left(\frac{\partial m_{yy}}{\partial x'} + \frac{\partial m_{xx}}{\partial y'}\right), \tag{2.66}$$

$$\vartheta = \arcsin \frac{p(1+\lambda)}{2}\left(\frac{\partial m_{yy}}{\partial x'} - \frac{\partial m_{xx}}{\partial y'}\right).$$

Substituting for ϑ in eqns. (2.54) and 2.55)

$$\left. \begin{aligned} \varepsilon_x &= \left(\left\{1 - \sqrt{\left[(1+\lambda)^{-2} - \frac{p^2}{4}\left(\frac{\partial m_{yy}}{\partial x'} - \frac{\partial m_{xx}}{\partial y'}\right)^2\right]}\right\} + p\frac{\partial m_{xx}}{\partial x'}\right), \\ \varepsilon_y &= \left(\left\{1 - \sqrt{\left[(1+\lambda)^{-2} - \frac{p^2}{4}\left(\frac{\partial m_{yy}}{\partial x'} - \frac{\partial m_{xx}}{\partial y'}\right)^2\right]}\right\} + p\frac{\partial m_{yy}}{\partial y'}\right). \end{aligned} \right\} \tag{2.67}$$

Thus the three components of strain may be evaluated from relations (2.66) and (2.67) by using the moiré data $\partial m_{xx}/\partial x'$, $\partial m_{xx}/\partial y'$, $\partial m_{yy}/\partial x'$ and $\partial m_{yy}/\partial y'$ taken at the point.

The terms in braces in eqns. (2.67) account for the moiré fringes formed by linear disparity and angular disparity of the reference grating, as well as for rigid body rotation of the specimen grating. These terms may be considered as correction terms added to the basic terms given by the pitch multiplied by the partial derivatives of fringe orders with respect to x'- and y'-axes respectively. The shear strains are directly and simply evaluated from the sum of the partial derivatives of moiré fringe orders with respect to x'- and y'-axes multiplied by the pitch. There are neither correction terms for the shear components due to linear or angular disparity of the reference grating, nor for rigid body rotation of the specimen.

The correction terms do not require separate measurements for their evaluation but merely make use of fringe gradients already measured for shear strain calculations.

The method as it has been developed is suitable for small-strain fields. It is easy to extend the method to cases of large strains. A discussion by Parks,[23] on the basic paper by Post, gives the general features for extending the grid-analyser method to large-strain fields.

The main advantage of the method is that it yields from one photograph all the necessary information for the evaluation of the strain components everywhere on the strain field. Since displacements of the initial pattern exceed those introduced by the strain field, the moiré fringes corresponding to the total displacement field exhibit an orderly distribution and are closely spaced. Therefore their interpretation is easy. For λ-negative, the fringe m_{xx} and m_{yy} increase in the positive x- and y-directions. Conversely, for λ-positive, m_{xx} and m_{yy} decrease in the positive x- and y-directions respectively. Therefore, any uncertainty in assignment of fringe orders is avoided in this method. When the moiré fringe orders are established along the x- and y-directions throughout the field, their gradients and the strains are evaluated at every point.

The only drawback of the method is the necessity of lengthy pointwise calculations for evaluating each component of strain at each point of the field, but this difficulty may be overcome by the use of a digital computer.

In the special case in which rigid-body rotations are negligibly small, the term under the square root in eqns. (2.67) becomes a constant. The constant term may be evaluated at some point of the strain field and used in the calculations of normal strains everywhere in the field. This considerably reduces strain evaluations.

2.8 Fractional moiré fringes by the light intensity–displacement law

In the previous discussion of moiré patterns a discontinuous relationship between the displacement field and the fringes of moiré patterns was developed. Displacements which are equal to integral multiples of the pitch of the reference grating or half the reference grating pitch were related to the points of maximum and minimum light intensity respectively.

[23] Parks, V. J., *Exp. Mech.* **6** (5) 287 (1966).

In the application of moiré patterns to strain analysis the most important restriction is the fact that the information extracted from moiré fringes is limited to discrete values. This restriction can be eliminated by the introduction of the light intensity–displacement law.

For the extraction of information from a moiré pattern, two distinct steps are existing, that is the recording of the moiré pattern on the photographic film or plate and the reading of the displacement information from the moiré pattern.

Either the location of the intensity maxima and minima of the moiré fringes or the measurement of coordinates of location of these points to a fixed coordinate system are carried out by the eye, which operates as an intensity sensing device.

The contrast γ_e between two points A and B having light intensities I_A and I_B respectively $(I_A > I_B)$ is given by

$$\gamma_e = \frac{I_A - I_B}{I_A}.$$

The ability of a normal eye to separate two adjacent points of different intensities is limited to values $\gamma_e \geqslant 0\cdot02$.[24] For a sinusoidal intensity distribution at regions near the extrema the contrast is

$$\gamma_e = (1 - \cos \alpha)$$

where α is the angle of deviation of the second point from the neighbouring extremum. Introducing the value of $\gamma_e = 0\cdot02$ the angle α becomes

$$\alpha = \pm 11^\circ,$$

which represents an error $e = \dfrac{11}{360} = 3$ per cent. This error limits the sensitivity of the method so that the smallest increment of displacement Δu which can be measured is

$$\Delta u = 0\cdot03p.$$

[24] Born, A., and Wolf, E., *Principles of Optics*, Pergamon Press, 2nd edn., 1964, p. 184.

The above error may be further increased in moiré patterns with an uneven illumination where at the densest regions of fringes this error is raised to 20 per cent. One remedy is to apply the differential method with the reference grating having different values of λ which displace the positions of densest fringe areas and allow an averaging process of errors.

The use of a photosensitive device whose electric output is linearly proportional to the light intensities of the image provides

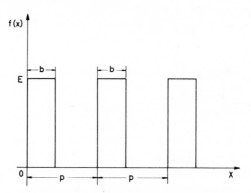

Fig. 2.22. Light intensity distribution emerging from an undeformed amplitude grating expressed by its transmission function $f(x)$.

a means of precisely locating points whose intensities are measured. This device is based on a light intensity–displacement law, which relates in a continuous manner both quantities.

The idea of transplanting the basic concepts of a continuous law yielding a measurement of relative signal distributions from communication theory and electronics is due to Sciammarella et al.,[25] who adapted the essential parts of this very fruitful theory to the

[25] Sciammarella, C. A., Exp. Mech. 5 (5) 154 (1965). See also: Ross, B. E., Sciammarella, C. A., and Sturgeon, D., Exp. Mech. 5 (6) 161 (1965). Sciammarella, C. A., and Sturgeon, D., Exp. Mech. 6 (5) 235 (1966). Sciammarella, C. A., Proc. Second Int. Cong. Exp. Mech. (Rossi, B. E., Ed.), Soc. Exp. Stress Analysis, Westport, Conn., 1966, p. 62.

continuous measurement of displacement components in a strain field by the moiré patterns.

The light intensity distribution emerging from a grating is of the form shown in Fig. 2.22. Mathematically it can be expressed by a transmission function $f(x,y)$ expressed as

$$f(x,y) = \frac{\Phi_0^z(x,y)}{\Phi_0^l(x,y)},$$

where Φ_0^z and Φ_0^l are the disturbances reaching and leaving the gratings. Since moiré patterns formed from interference of coarse gratings are concerned with amplitude gratings which have a real mathematical expression, the following theory will be restricted to this type of transmission functions.

If two amplitude gratings, the reference and the specimen gratings illuminated under predominantly incoherent illumination, are superposed face to face, the resulting transmission function $f_T(x,y)$ is given by

$$f_T(x,y) = f_R(x,y) . f_S(x,y), \tag{2.68}$$

where $f_R(x,y)$ and $f_S(x,y)$ are the transmission functions of RG and SG respectively.

If the pitches of the two gratings are different due to a linear and an angular disparity, the resulting transmission function will contain, besides the images of the rulings of the grating, a beat phenomenon forming the moiré fringes. In the application of moiré fringes total intensities are disregarded, while the average intensities yielding the moiré fringes are of main interest.

Consider an amplitude type grating (Fig. 2.23). The expansion of its transmission function in a Fourier series is

$$f(x) = C_0 + 2 \sum_{n=1}^{\infty} C_n \cos \frac{2n\pi}{p}, \tag{2.69}$$

where p is the pitch of the grating and the constants are

$$C_0 = \frac{Eb}{p}, \quad C_n = C_0 \frac{\sin (n\pi b/p)}{n\pi b/p}, \tag{2.70}$$

where b/p is the transmittance of the gratings (usually $b/p = \frac{1}{2}$; Fig. 2.22) and E the amplitude of the mean light intensity of the initial rectangular pulse.

If the reference grating is assumed to have a linear disparity $(p_r = p_s(1+\lambda))$ and an angular disparity ϑ and if normalized co-

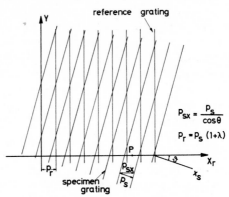

Fig. 2.23. Geometry of two line gratings presenting a linear and an angular disparity.

ordinates ξ_s and ξ_r of a generic point P of the specimen related to the pitches p_s and p_r (Fig. 2.23) are defined by

$$\xi_r = \frac{x}{p_r} \quad \text{and} \quad \xi_s = \frac{x \cos \vartheta}{p_s},$$

the fractional displacement ρ of the one grating with respect to the other is given by

$$\rho = (\xi_r - \xi_s), \tag{2.71}$$

where ρ increases by an integer along the maximum of moiré fringes. For 50 per cent transmittance of the gratings the equation of transmission becomes

$$f(\xi) = C_0 + \frac{4C_0}{\pi} \sum_{n=0}^{\infty} \frac{\cos (2n+1)2\pi\xi}{(2n+1)}. \tag{2.72}$$

Since moiré fringe formation is related to average intensities, it is necessary to introduce an averaging process through a cross-correlation function defined as

$$\bar{f}_T(x',y') = \lim_{A \to \infty} \frac{1}{A} \int_A f_s(x,y) f_r((x+x'),(y+y'))dA, \qquad (2.73)$$

where $\bar{f}_T(x',y')$ is the resulting average transmission function, x, y is a coordinate system fixed to the gratings, x', y' are relative displacements independent of x and y and A is the area of integration, which in the case of gratings may have as limit the infinity.

In order to evaluate the average transmission function $\bar{f}_T(x',y')$ it is necessary to introduce the Fourier transforms of both sides of eqn. (2.68) which yield

$$\bar{F}_T(g,h) = F_s(-g,-h) . F_r(g,h), \qquad (2.74)$$

where F designates the Fourier transforms of the functions in relation (2.68). Since the average intensity function is sought along a line parallel to the x-axis the second variable h in the transform space may be excluded from the above relation which becomes

$$\bar{F}_T(g) = F_s(-g) . F_r(g). \qquad (2.75)$$

Equation (2.75) implies that each component of the resulting transmission function, which is assumed as being composed by the sum of space harmonics of all possible frequencies, derives from the corresponding components of the transmission functions $f_s(x,y)$ and $f_r(x,y)$. This means that the superposition of two gratings creates a linear filter effect.

Each process grating may be considered as formed by the sum of sine-wave gratings of different frequencies and amplitudes suitably superimposed to satisfy eqn. (2.72). It is possible, in computing the transform of eqn. (2.72), to separately deal with each sine term and add the corresponding results. In the case where there is an infinitesimal linear and angular disparity in pitches of the two process gratings, and therefore the moiré field is homogeneous, it is easy to obtain the transform function for each term of each transmission function of the series by using relation

$$F(g) = \lim_{L \to \infty} \frac{1}{2L} \int_{-L}^{+L} f(x)e^{-2\pi igx}dx. \tag{2.76}$$

The transform terms of each transmission function introduced into eqn. (2.72) yield the transform terms of the cross-correlation function $\bar{f}_T$. If these terms are inverted and added, they yield the complete relation for $\bar{f}_T(\rho)$

$$\bar{f}_T(\rho) = C_0^2 + \frac{16C_0^2}{\pi^2} \sum_{n=0}^{\infty} \frac{\cos 2\pi(2n+1) \cdot x/f}{(2n+1)^2}, \tag{2.77}$$

where

$$f = \frac{p_r p_s}{p_r - p_s \cos \vartheta}.$$

Equation (2.77) shows that the transmission function $\bar{f}_T(\rho)$ corresponds to a positive lens for the gratings yielding an enlarged image of them in which the pitch p_s is replaced by f.

For an heterogeneous field the concept of local pitch must be introduced at a generic point x. The pitch may be regarded as a continuous variable with a very small change rate. The idea of the average pitch is also useful. It is defined as the length of the ruled area measured along a principal direction divided by the total number of rulings and it is called $\bar{p}$. Then the local pitch

$$p(x) = \bar{p} + \Delta p(x). \tag{2.78}$$

Likewise, a local angular displacement $\vartheta(x)$ and an average angular displacement $\bar{\vartheta}$ are defined and related by

$$\vartheta(x) = \bar{\vartheta} + \Delta\vartheta(x). \tag{2.79}$$

Introducing these definitions in the previous analysis for the homogeneous field and following the same steps, it can be shown by a similar process that for the heterogeneous moiré field the transmission function $\bar{f}_T[\rho(x)]$ may be expressed as

$$\bar{f}_T[\rho(x)] = C_0^2 + \frac{16C_0^2 K}{\pi} \sum_{n=0}^{\infty} \frac{\cos(2n+1)2\pi\rho(x)}{(2n+1)^2}, \tag{2.80}$$

where $\rho(x)$ is the variable fractional displacement of the two gratings due to the deformation of the specimen grating.

In the above-mentioned analysis the effect of the image-forming system on the moiré pattern is not considered. Following the same procedure it is possible to show that the transformation from the object to the image is a linear filter action as the action of interference of two coarse gratings.

The Fourier transform of the intensity distribution of the image[26] is given by

$$J(g,h) = F(g,h) \cdot L(g,h), \tag{2.81}$$

where $J(g,h)$ is the Fourier transform of the intensity function, $F(g,h)$ is the Fourier transform of the transmission function, and $L(g,h)$ is the transform function defined by

$$L(g,h) = \int\limits_{-\infty}^{\infty} \int\limits_{-\infty}^{\infty} G((g+g'),(h+h'))G^*(g',h')dg'dh', \tag{2.82}$$

where $G(g,h)$ is the so-called pupil function of the image forming system and the asterisk indicates the conjugate function. Equation (2.82) expresses the autocorrelation of the pupil function.

For an area of the exit pupil equal to a circle $A_p = \pi r^2$ the integral in eqn. (2.82) does not vanish in the common area of the two circles (Fig. 2.24) A and A' (hatched area), where the circle A' is similar to A but with its origin displaced by $-g$, $-h$.

The values of the function $L(g,h)$ are given as

$$L(0) = \pi\alpha^2 \quad \text{and} \quad L(h) = 2\alpha^2(\beta - \frac{1}{2}\sin 2\beta), \tag{2.83}$$

where α is equal to the radius of the pupil r divided by the wavelength of the light transmitted λ and the radius R at the Gaussian sphere, and $\cos \beta$ is equal to the distance h of centres of circles A and A' divided by 2α.

[26] Flugge, S. (Ed.), *Encyclopedia of Physics*, Vol. 24, *Fundamentals of Optics*, Françon, M., Interference, Diffraction, Polarisation, 1956, p. 344.

By applying eqn. (2.81) to relation (2.80) the intensity distribution in the image plane is given by

$$I(x) = C_0^2 L(0) + \frac{16 C_0^2 K}{\pi^2} \sum_{n=0}^{\infty} \frac{L(2n+1) \cos (2n+1) 2\pi\rho(x)}{(2n+1)^2}, \quad (2.84)$$

where K is a factor smaller than unity which takes into consideration the effect of the spread of the fundamental frequencies around the peaks of the discrete pulses.

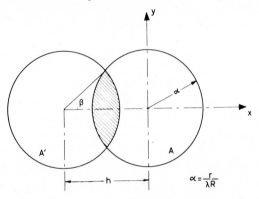

Fig. 2.24. Area of integration (hatched area) for the determination of the auto-correlation function related to the pupil function of the optical system.

Relation (2.84) shows that the intensities of higher frequencies monotonically decrease to zero with increasing frequency.

In the previous analysis it was assumed that the optical system was free of aberrations and perfectly focused. The presence of aberrations and or the lack of focusing considerably contribute to a decrease of the response of the lens system to higher frequencies. Therefore, the higher harmonics of the moiré system are attenuated and extinguished long before reaching the theoretical limit for the aperture system, which is expressed by the relation $h < 2r$. Consequently, if the numerical aperture of the system is reduced, it may filter out all higher harmonics than the first one. For gratings having a frequency of 300–500 lines per inch it is not necessary to take special precautions for the optical system because under ordinary conditions of formation and observation of the moiré fringes the condition of filtering the harmonics of higher order than the first are commonly met. Therefore the intensity distribution in the image will be given by

$$I(x) = C_0^2 L(0) + \frac{16 C_0^2 K}{\pi} L(1) \cos 2\pi\rho(x). \qquad (2.85)$$

This equation may be written for the general case as:

$$I(x) = I_0 + I_1 \cos 2\pi\rho(x), \qquad (2.86)$$

where I_0 is the average background intensity and I_1 is the intensity amplitude of the first harmonic. Both intensities are functions of the transmission functions of the gratings and the optical image forming system.

According to the definition of relative displacement $\rho(x)$ the intensity maxima are the loci of the projections of the displacements in the principal direction. These are equal to an integral number of the reference pitch, while the intensity minima are the loci where these projections are equal to an impair number of half the pitch of the reference grating.

The importance of eqn. (2.86) lies on the fact that it provides a continuous relationship between light intensity at a generic point and its displacement. It may be deduced from eqn. (2.86) that the displacement u in the x-direction is

$$u = \left(\frac{1}{2\pi} \arccos \frac{I_1 - I_0}{I_1} \right) \cdot p_r. \qquad (2.87)$$

The validity of eqn. (2.87) is extended to both homogeneous and heterogeneous fields. The increase of the sensitivity of the moiré method is no more based on the increase of the frequency of gratings since by eqn. (2.87) it is possible to measure fractions of the pitch of the gratings.

Since the required information in the moiré pattern is contained in the argument of the fractional displacement $\rho(x)$ one harmonic

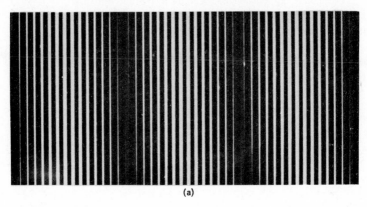

(a)

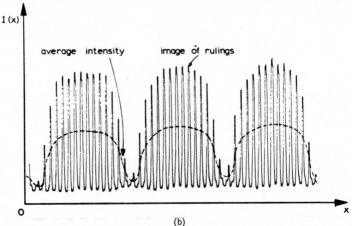

(b)

Fig. 2.25. Microdensitogram (b) presenting the structure of moiré fringes by a linear disparity of two coarse gratings (a). (Reference and specimen gratings coinciding with the object plane.)

suffices to determine this argument, and no information is lost when working with the truncated relation (2.85).

In the discrete moiré law, which is applied in the usual cases of moiré patterns, it is preferred to work with points of maximum and minimum intensity since these are easily detected by eye. If a photosensitive device is used, capable of measuring relative intensities, it is possible to extend the measurements to all fractional points of the intensity curve since all the points of the field contain the same information.

The physical evidence corresponding to eqn. (2.84) is shown in Fig. 2.25 where the record of moiré fringes is depicted as obtained

FIG. 2.26. Photograph of a microdensitometer.

by a microdensitometer. The moiré pattern corresponds to a tensile specimen having a specimen grating of a frequency of 25 lines per inch.

An automatic recording microdensitometer[27] may be used for recording the light intensity distribution in a moiré pattern. The instrument has a dual function, i.e. as an optical density sensing device, as well as an instrument which measures relative spacing

[27] Joyce, Loebl & Co. Ltd., Automatic Recording Microdensitometer, Model MK IIIc (Princeway, Team Valley, Gateshead 11, England).

between fringes and thus defines the location of points whose density was measured with respect to a fixed coordinate system (Fig. 2.26).

The principle of operation is based on a true double-beam light system in which two beams from a single light source are alternatively switched to a single photomultiplier. When the two beams

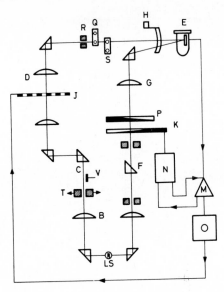

FIG. 2.27. Schematic representation of a microdensitometer.

are of a different intensity, a signal produced by the photomultiplier causes a servomotor to move an optical attenuator so as to reduce the intensity difference to zero. In this way a continuously null balancing system is obtained in which the position of the optical attenuator is recording the density of any particular point of the specimen.

The principle of the instrument is schematically illustrated in Fig. 2.27. A filament light source LS yields two beams of light. One beam follows the $BCDE$ path, while the second follows the

FGE path. At the position *H* a synchronous motor equipped with a shutter alternately exposes either beam on to the photo-multiplier *E*. The position of the specimen is at *J*, while at *K* an optical grey wedge is placed. The position of the optical wedge, to which a writing pen is directly attached, is controlled in the following manner. The signal from the photomultiplier *E* is fed to the amplifier *M*, the output of which is applied to the servo-motor *N*. Servostability and sensitivity control are achieved by applying a feedback signal derived from the tachogenerator, coupled to the servomotor *N*, to the amplifier. The specimen and record tables are driven by another servosystem *O* at a speed proportional to the rate of change of density. The position of the optical attenuator and, therefore, of the pen is controlled by a manually operated wedge *P* in the light path *F*, *G*. Neutral filters *Q* are provided to control the intensity of the specimen light path *BCDE* in order to obtain balanced conditions with wide aperture openings *R* as the reference light path intensity is set to work within photomultiplier's *E* range under all conditions. The principle of the true double-beam light system used in the instrument has the advantage of making the instrument almost independent of its own parameters, and complete reproducibility of record is possible. Then the instrument records density directly and linearly within the range of densities specified.

The light intensity $I(x)$ can be obtained from the photographic negative of the moiré pattern. The incident exposure $E(x)$ on the film is retrieved in the instrument from the density trace $D(x)$. The relation connecting $E(x)$ and $D(x)$ is

$$D(x) = I(x) \cdot t = \gamma \log E(x), \qquad (2.88)$$

where γ is the slope of the so-called H and D characteristic curve of the film and t is the exposure time. In order to have a unique value of γ all over the negative, it is advisable to work at the straight-line portion of the H and D characteristic curve $[D(x) = f \log E(x)]$, far from its knee and elbow, where γ is changing rapidly for various values of exposures. In the case of a moiré pattern the exposure is of a sine-wave form:

$$E(x) = E_o + E_a \cos \frac{2\pi x}{f}. \qquad (2.89)$$

The transmittance of the film T is defined as the ratio of the emerging light intensity I_e from the film and the incident light intensity I_i impinging on the film. The inverse of transmittance is called opacity O of the film and is given by

$$O(x) = [T(x)]^{-1} = \frac{I_i}{I_e}. \qquad (2.90)$$

Since the density of the film is expressed in a logarithmic scale, it is valid that

$$D(x) = \log O(x). \qquad (2.91)$$

Substituting eqns. (2.89) and (2.91) into relation (2.88) yields that

$$O(x) = \left(E_o + E_a \cos \frac{2\pi x}{f} \right)^{\gamma}. \qquad (2.92)$$

The value of γ depends on the type of film and the developing process. For accurate measurement of the relative opacities of a negative presenting a moiré pattern it is necessary to choose a film having a γ approximately equal to unity and to work in the straight line part of its Hurter and Driffield characteristic curve in order to have a constant value of γ all over the field. In this case a linear relation connects the measured opacity and sine-wave variation of exposure due to moiré effect.

The error introduced by a variation in the value of γ can be derived from eqn. (2.92) by differenting with respect to γ, i.e.

$$\frac{dO}{O} = \left[\ln E_o + \ln \left(1 + \frac{E_a}{E_o} \cos \frac{2\pi x}{f} \right) \right] d\gamma,$$

which, expanded in a Taylor series, yields

$$\frac{dO}{O} = \left[\ln E_o - \sum_{n=1}^{\infty} \frac{(-1)^n}{n} \left(\frac{E_a}{E_o} \right)^n \cos^n \frac{2\pi x}{f} \right] d\gamma. \qquad (2.93)$$

It may be deduced from eqn. (2.93) that changes in values of γ

introduce high-order harmonics which affect the sinusoidal form of the intensity curves by sharpening their hills and flattening their troughs. The intensity distribution across the moiré fringes, as it is recorded from the microdensitometer, is a means of detecting any departure from the ideal distribution due to improper values of γ (Fig. 2.25).

The above analysis is only valid for an ideal system relating displacements to light densities. However, the continuous trace of the incident exposure contains, besides the phase modulated spatial curve yielding the information on the relative displacements along a given line of the specimen, spurious amplitude and phase variations due to noise. It is beyond the scope of this section to present a complete analysis of errors in moiré patterns. The type of errors due to imperfections of gratings will be discussed in a separate chapter. In this section only errors which influence the accuracy of evaluation of fractional orders will be mentioned. The aim is to free relation (2.93) from the noise in which it is merged before proceeding to the determination of the displacements.

These particular sources of error influencing the accuracy of the values of fractional fringes may be classified into three types: (a) errors due to change in the background intensity, (b) errors due to change in the intensity of moiré fringes, and (c) errors due to local disturbances.

Changes in background intensity are created by (a) an uneven illumination originating at the light source, (b) changes in the transmission properties of transparent materials or changes in the reflection properties of opaque reflecting surfaces, and (c) uneven intensity distribution at the boundaries of the image plane due to the optical properties of the lens system.

While the first cause of changes in background intensity is susceptible to a simple correction, the other two causes necessitate a complex optical system for their correction. These are long wavelength variations. The most frequent causes of short wavelength variations are the images of the reference and specimen rulings on the moiré patterns, scratches, dust and pits on the surface of the specimen. These local effects tend to be random and they contribute only to the fine details of the function. Since the photo-optical system used for the formation of moiré fringes is consisting of a light source, lenses, gratings and an angular aperture, it has inherent low-pass filter capabilities and yields a means of eliminating short wavelength variations.

On the contrary, long wavelength variations due to changes in background intensity, as well as due to change in the amplitude intensity of fringes, contain a large number of Fourier components, some of which may eventually have frequencies close or overlapping to the signal frequency. This fact makes the elimination of long wavelength variations difficult.

One of many other possible ways of removing both long wavelength and amplitude perturbations from the signal is the use of digital bandpass filters, as they have been successfully used by Sciammarella et al.[28] In the process the problem was handled as a case of narrowband frequency modulation. In this case the carrier frequency must be of the same order of magnitude of the highest significant modulation frequency. The carrier is an initial moiré pattern formed by a disparity in pitch of the two gratings. By estimating the initial interfringe spacing f_i, as well as its variation limits $\pm \Delta f_i$, due to deformation of the specimen, a suitable electronic filter can be used to eliminate the noise since, during the scanning process, space frequencies are transformed into time frequencies.

The initial steps in the moiré data process are the following: the signal from the microdensitometer, after being filtered and amplified, is fed into an analog-to-digital converter, which stores the information in a punched tape. The punched-tape information is converted into the card format required by a digital computer, which is utilized further to perform the succeeding stages of data processing. Figure 2.28 shows schematically the basic steps of the data processing.

The intensity curve shown at the bottom of Fig. 2.28 is stored in digital form on the computer cards. Variations in amplitude and background illumination are shown in Fig. 2.29, which are

[28] Sciammarella, C. A., *op. cit.*, p. 68.

not completely removed by the filter, since the filtering process cannot remove the noise falling within the range of the signal.

To correct these remaining effects the computer performs the following operations:

(a) The whole intensity trace is scanned to determine the position of the extrema of the intensity curve.

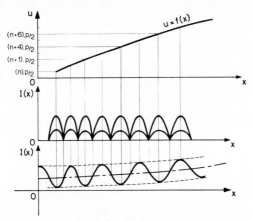

Fig. 2.28. Schematic presentation of succeeding steps performed by a digital computer during the interpolation process of moiré data.

(b) The values and positions of the mid-ordinates of each half-wavelength between extrema are evaluated.

(c) The intensity trace is rectified and the mean curve of the background intensity is translated to coincide with the distance axis.

(d) The absolute values of the data are taken so that the whole intensity curve is rectified.

(e) The half-cycle loops of varying amplitude are normalized.

With the normalization process the data are ready for the final computation of the displacements for which the computer makes the following steps.

(f) The computer finds for each point corresponding to a fractional fringe order the value of the arc corresponding to the ordinate of the point.

(a)

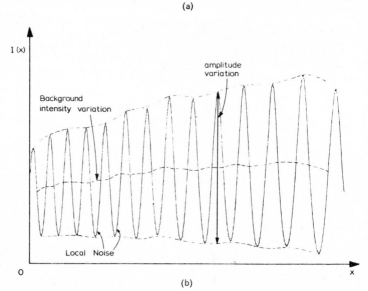

(b)

FIG. 2.29. Moiré pattern (a) and corresponding microdensitogram (b) showing three sources of error, i.e. local noise, background intensity variations and intensity amplitude variations.

(g) The intensity ordinates of the points are added pointwise to evaluate their displacements.

The computer may be eventually ordered to execute a final step of graphical or numerical differentiation of the displacement curve along the x- and y-axes to yield the direct and cross derivatives of the displacement components.

The curve of displacements which the computer yields may show some remaining noise, but the scatter is not generally significant. Since differentiating a curve is a diverging process a small scatter in the displacement curve may introduce large errors in computing strains. It is, therefore, necessary to introduce a smoothing process in the displacement curve in order to avoid large scatter in strains. The whole problem of computing strains from displacement data rests on the possibility of finding a successful technique of numerical differentiation.

2.9 Fractional moiré fringes by isodensitracing

In the preceding section the mechanism of creation of moiré signal was analysed and the process of storing the information contained on photographic negative, with all subsequent changes of the signal as well as the mechanism of information retrieval, was developed. In this process the moiré signal was separated from the noise and the information contained in the moiré signal was retrieved by the use of an electronic filter and an analog-to-digital converter, which stored all information on a punched tape. The punched-tape information was converted into the card format required for a digital computer and the remaining steps of information retrieval was carried out in digital form.

While this technique is efficient in solving the problem of increasing the sensitivity of moiré method without necessitating an increase in the line frequency of the gratings, it presents the disadvantages:

(i) While the coarse moiré pattern yields a whole field of view of the displacement distribution, the interpolation technique becomes a linewise process necessitating a series of traces over successive lines in the specimen in order to yield a detailed picture of the displacement distribution.

(ii) The interpolation process is a complicated process consisting of a series of numerous steps, which are time consuming, necessitate elaborate equipment, skilful techniques for the accurate evaluation of the data, and each of the steps introduces its inherent contribution of errors in the whole process.

(iii) The real intensity signal obtained from the moiré pattern is expressed by a series of sinusoidal terms of increasing frequency. These series are truncated to the first term. The higher harmonics are considered as contributing an insignificant amount to the intensity variation by the filtering action of the photo-optical system. While this may be generally a good approximation for stringent conditions in the form of gratings and in the photo-optical arrangement used for recording the moiré fringes, in some cases where the relative displacement function is rapidly varying between successive moiré fringes, the amplitude coefficients of the higher harmonics in the Fourier decomposition of the signal image may persist with progressive vigour to their damping and, therefore, they significantly contribute to the periodic form of the intensity curve. These terms, if neglected in the interpolation process, may introduce significant errors.

Another efficient way of obtaining fractional moiré fringes and retrieving the necessary information from a moiré pattern by applying the continuous light intensity displacement law is by optical means by using an isodensitracer. This method was introduced by Theocaris.[29]

The moiré pattern formed by two superposed amplitude gratings RG and SG, the one of which has undergone an angular displacement ϑ and/or some deformation along its principal direction due to an applied strain field, presents an intensity variation along the

[29] Theocaris, P. S., *Quart. Sci. Rev.* **36**, 626 (1967).

traverse xx' (which for simplicity is taken to coincide with the principal direction of RG) given by:

$$I(x) = I_0 + \sum_{n=1}^{\infty} I_n \cos 2n\pi\rho(x), \qquad (2.94)$$

where I_0 is the average background intensity and I_n are intensity coefficients of higher harmonics. The quantity $\rho(x)$ expresses the component of the fractional displacement of SG with respect to RG along the traverse xx'. The displacement $\rho(x)$ is given by

$$\rho(x) = (\xi_r - \xi_s), \quad \text{where} \quad \xi_r = x/p_r, \quad \xi_s = \frac{x}{p_s/\cos\vartheta} \quad \text{and} \quad p_s = p_r(1+\lambda),$$

λ being an infinitesimal positive or negative quantity expressing the linear disparity of the line pitches p_r and p_s of the gratings. The fractional displacement $\rho(x)$ is a continuous function, the integral values of which correspond to the middle points of dark fringes while the half values coincide with the middle points of bright fringes.

The fractional displacement $\rho(x)$ contains all the information which is required from a moiré pattern to evaluate the components of displacement along the principal direction of the gratings. The argument $\rho(x)$ of the function $I(x)$ may be determined by the first harmonic given by eqn. (2.94) in the case where the coefficients of all but the first harmonic are of small amplitude and produce insignificant effects. However, this is a seldom case in phase gratings which are specially blazed for the wavelength of light used in the direction of observation. Generally, many terms have coefficients of sufficiently similar amplitude which create significant modulations to the basic harmonic and must be taken into consideration for the evaluation of $\rho(x)$.

Amplitude gratings, when viewed without the interposition of an optical element, present intensity distributions with many harmonics having coefficients of similar intensity to the intensity of basic harmonic and create the intensity distributions of the sawtooth form.

Let us consider two line gratings superposed so that their rulings subtend an angle ϑ. The pitches of the gratings are p_r and p_s and

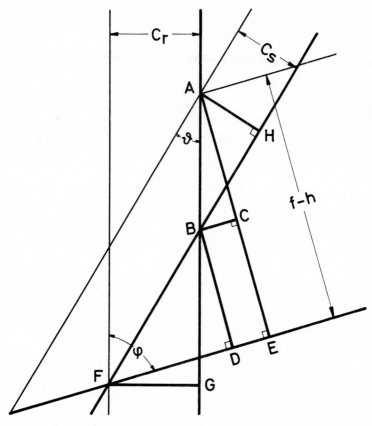

FIG. 2.23. Geometry at the vicinity of the intersection of the boundaries of two generic opaque bars in two superposed gratings subtending an angle ϑ.

the ratios of the open slit to the opaque bar widths are $t_r = c_r/b_r$ and $t_s = c_s/b_s$ respectively. While the interfringe spacing f depends only on p_r, p_s and ϑ, the thickness h of the opaque moiré fringe depends also on the ratios t_r and t_s. Figure 2.23 illustrates the geometry at the vicinity of point A which lies on the intersection of the boundaries of two generic opaque bars. The distance AE which is equal to the width of the open moiré fringe is given by:[30]

$$f - h = AC + CE = c_s \frac{\sin \phi}{\sin \vartheta} + c_r \frac{\sin (\phi - \vartheta)}{\sin \vartheta}. \qquad (2.95)$$

But since it is valid that:

$$f = \frac{p_r}{\sin \vartheta} \sin (\phi - \vartheta) = \frac{p_s}{\sin \vartheta} \sin \phi$$

relation (2.95) becomes

$$\frac{h}{f} = \frac{b_r b_s - c_r c_s}{p_r p_s}. \qquad (2.96)$$

In the case where both gratings are identical ($b_r = b_s = b$, $c_r = c_s = c$ and $p_r = p_s = p$) the above relation reduces to

$$\frac{h}{f} = \frac{b - c}{p}. \qquad (2.97)$$

The width h given by eqn. (2.95) represents the width of the dark moiré fringe in which a complete light extinction occurs and it is different from the apparent fringe width which the eye sees.

In the case of gratings which present a 50 per cent transmittance, that is the width of the open slits is half the line pitch of the gratings (Fig. 2.24a, $b = c = p/2$), the coefficients I_n take values which create a symmetric saw-tooth curve so that the average intensity of the emergent light varies between a minimum intensity equal to zero and a maximum intensity equal to 50 per cent the unobstracted intensity of light. The width h of the fringes in this case becomes zero.

For complementary gratings, that is for gratings with inversely proportional ratios of clear slit to opaque bar widths, the moiré pattern formed has a trapezoid symmetric shape with minimum points of zero intensity and maxima of intensity equal to

[30] Zandman, F., Holister, G. S. and Brcic, V., *Jnl. Strain Analysis*, **1**, 1 (1965).

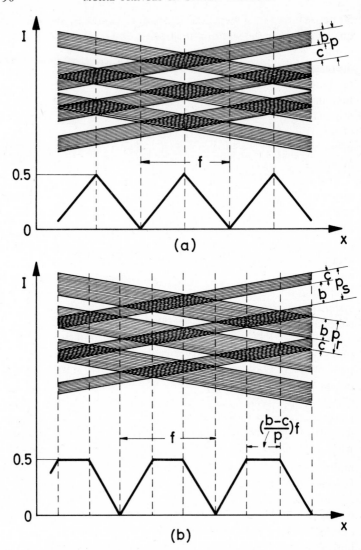

Fig. 2.24. Schematic representation of the moiré fringe widths formed by identical gratings having a 50 per cent transmittance (a) and by complementary gratings (b).

50 per cent which extend over a width $h_{max} = (b-c)/p$ where the quantities b, c and p belong to the more opaque grating and are given in Fig. 2.24b. The sharpness of moiré fringes increases with decreasing transmittance of the more opaque grating. Sharpening of moiré fringes can be achieved with non-complementary gratings, but, in this case, the intensity difference between maxima and minima of the moiré pattern is drastically reduced. This reduction of intensity in moiré fringes unfavourably influences the contrast of moiré fringes and therefore this type of grating pairs must be avoided.

The moiré signal formed by two superposed gratings which was previously described corresponds to the ideal case of perfect gratings directly observed without the interposition of an image-forming device. Moreover, it was assumed that the illumination of the gratings was diffuse and uniformly distributed. This ideal arrangement cannot generally be met in experiments and there are always deviations from the ideal situation due to the imperfections in illumination, defects of gratings and the existence of the optical system used to record the moiré pattern. These deviations may be either amplitude or phase deviations.

The amplitude deviations may result from an uneven illumination and transmittance or reflectance of the specimen, from variations of the ratio c/p, i.e. the open-slit width to the line pitch ratio, to which the transmittance of the grating is directly proportional, from an eventual gradual variation in amplitude of the higher harmonics in the series of eqn. (2.94) expressing the intensity $I(x)$, from the variation of frequency of moiré fringes over the strain field and from the choice of the type of film used for copying the moiré pattern, the exposure time and the developing process. Indeed, an appropriate choice of the three latter factors is essential for the faithful reproduction on the film of the undistorted image of the moiré pattern. This can be achieved if the maxima and minima in the optical density of the film lie on the straight-line portion of the characteristic curve of the film which expresses the optical density D as a function of the logarithm of the exposure (E).

The phase variations may be divided into short wavelength

variations and long wavelength variations. The short wavelength variations are due to the images of the rulings of the gratings on the moiré fringes and to eventual scratches, pits and dust gathered on the surface of the specimen. The long wavelength variations are caused by an uneven illumination combined with variations in the transmittance or reflectance of the specimen.

The image-forming device consisting of the camera lens and the film, which is used to record the moiré pattern, presents a low-pass filtering action of spatial frequencies.[31] Since either element of the system individually presents this capability, the combination constitutes a system operating as two filters in series. The filtering effect can be considered as the process by which a stop is intervened by the optical element to higher spatial harmonics of the Fourier decomposition of the signal, which are unable to appear on to the recorded image. The filtering effect is associated with a degrading effect on the image since the infinite expansion that describes all the details of variation of the signal is truncated after a certain finite number of terms in its Fourier expansion. Besides, the filtering effect influences the remaining terms of the Fourier expansion of the signal by damping the corresponding amplitude coefficients with progressive vigour in higher harmonics. These degrading and damping actions of the filter create an increasing reduction of contrast for higher spatial frequencies. A further degradation of the light signal occurs by the aberrations of the lens and by diffraction, which was not previously mentioned.

The film has a similar filtering effect on the signal. The damping of the intensity coefficients of higher harmonics by the film is achieved by scattering the light within the emulsion layer. The degradation of contrast of the film is increased with increasing the spatial frequency of the object. Each film possesses a particular characteristic curve yielding this degradation in contrast and films have different limiting values of maximum resolution.[32]

Thus, the combination of the lens with the appropriately selected

[31] O'Neill, E. L., *I.R.E. Transactions on Information Theory*, **IT-2,** 56 (1956).
[32] Kodak Pamphlet P-49 (49A), Modulation Transfer Data for Kodak Films, Eastman Kodak Co., Rochester, N.Y., U.S.A., 1962.

film for recording the moiré pattern creates an image of the pattern which suffers, besides a reduction of its contrast, a distortion of its original shape by rounding the edges of the saw-tooth or trapezoidal signal emitted by the two superposed gratings. This rounding of the edges of the signal is the result of the truncation of the Fourier series expansion of the original signal and of the progressive damping of the intensity coefficients of the higher harmonics. However, the evaluation of the variation of the fractional displacement, which yields the strain distribution and constitutes the variable argument of the function $I(x)$, is independent of the type of the carrier signal and its modification due to filtering provided that this modification does not influence its modulation due to the application of the displacement field. The only difference is that in the case when the carrier signal is sinusoidal the extraction of information from it is much easier than with a carrier signal of an arbitrary form.

While the image-forming device presents by itself an inherent property to filter out high-frequency noise there are also some adjustments of the optical system which yield additional modification of the transfer characteristics of the image-forming system.

It is possible to damp high harmonics in a signal by decreasing the aperture of the diaphragm of the camera lens.

For a lens the complex amplitude distribution is positive and less than unity over the entire area of the aperture and zero outside this domain. Therefore the resolution of a lens depends on the diameter D of its diaphragm and it is given by the ratio D/λ for the wavelength λ of the light used. Thus, by changing the size of the diaphragm of the camera lens it is possible to filter out and damp further intensity coefficients of high harmonics, which were unaffected by the inherent filtering action of the image-forming device.

However, the method to reduce the ratio of the noise level to signal level by diminishing the size of the aperture has some limitations in moiré applications since it necessitates long exposure times to photograph the moiré pattern.

Another possibility to filter out higher spatial harmonics and to reduce the ratio of noise level to signal level is by defocusing the camera lens. The unit intensity response h of the camera lens is given by:[33]

$$h(r) = \frac{I}{a}\left(\frac{Z_o}{Z_i}\right)^2 \qquad (2.98)$$

where I is the intensity of the point source, Z_o and Z_i are the distances of the object and image respectively and a is the distance of defocusing. The above relation is valid for a circle defined by a radius r given by the absolute value of the distance of defocusing divided by Z_i. Outside this circle the value of h rapidly tends to zero.

Relation (2.98) suggests another means of filtering out high harmonics of the signal by defocusing the lens of the image-forming device. For a lens the f-number of which is selected to yield a minimum of diffraction and aberration effects the depth of focus which corresponds to the limit of sharpness of the image is given by:[34]

$$\delta D \approx 0 \cdot 0025 \, (f\text{-number})^2 \, \text{(mm)}$$

This relation implies that the permissible variation in the position of the copying device from its exact focus is very limited and a small tilting of the film-plane to the optical axis of the camera may create over the image plane zones of different sharpness and resolution of the reproduced signal. Therefore, a small parallel displacement of the film from its exact focus position results in an efficient filtering of high spatial harmonics of the signal.

All these methods are perfectly permissible and suitable to filter out and to conveniently damp all spurious high spatial frequency disturbances in the signal since they only affect the high frequency harmonics of the carrier intensity signal and they do not affect its modulation due to an applied displacement field. However, great care must be taken in the copying characteristics of the system since the intensity distribution of the moiré signal

[33] Cheatham, T. P. and Kohlenberg, A., *I.R.E. Convention Record*, Part 4, 6 (1954).

[34] Kodak Pamphlet P-52, Techniques of Microphotography, Eastman Kodak Co., Rochester, N.Y., U.S.A., 1963.

by isodensitracing is retrieved from the density trace of the photographic negative of the moiré pattern. An amplitude deviation of the copy from the moiré signal, while it does not influence the positions of the entire and half orders of fringes, may displace the positions of the fractional orders defined by isodensitracing.

The relation connecting the incident exposure $E(x)$ along any traverse xx of the negative is given by eqn. (2.88). For the accurate measurement of the relative opacities of a negative depicting a moiré pattern it is essential to establish a constant value of γ in order to have a linear relationship between density and exposure. This can be achieved if the type of film, the exposure time and the developing process are chosen such that the value of γ is restricted in the linear portion of the H and D characteristic curve, i.e. far from its knee and elbow, where γ is changing rapidly for various values of exposure. In the case of a moiré pattern obeying relation (2.94) the exposure is of the form:

$$E(x) = E_o + \sum_{n=1}^{\infty} E_n \cos \frac{2\pi n x}{\rho(x)} \qquad (2.99)$$

Introducing eq. (2.99) into eq. (2.88) and using the expression for the opacity $0(x)$ of the film defined by eqns. (2.90) and (2.91) it can be readily deduced that:

$$0(x) = \left[E_o + \sum_{n=1}^{\infty} E_n \cos \frac{2\pi n x}{\rho(x)} \right]^{\gamma} \qquad (2.100)$$

For $\gamma = 1$ there is a linear relationship between opacity and the sinusoidal form of exposure as it may be easily derived from eqn. (2.100). Therefore, a measurement by the isodensitracer of the densities of the photographic negative in a logarithmic scale directly yields the opacity and exposure variations of the film which faithfully reproduce the contrast of the moiré signal.

An isodensitracer was used for the retrieval of the signal which was faithfully recorded on the negative film. The instrument is a high-speed direct reading isophotometer designed for contour mapping of two-dimensional photometric information. It automatically scans and measures densities all over the film transparency, plots contour curves of equal density and, therefore, yields the exact positions of the entire and half order moiré fringes, as well as of fractional-order moiré fringes. The instrument has a magnification ratio range from the negative to the record between 1:1 and 2,000:1. It can scan either a large field or a detail from

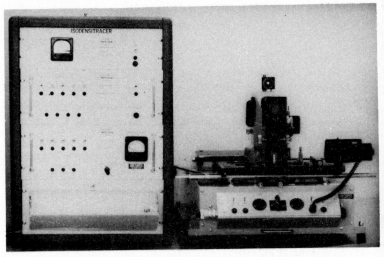

FIG. 2.25. Photograph of an isodensitracer consisting of a microdensitometer with an isophotometer attachment and a programmer.

the whole negative transparency and possesses two density ranges covering densities between zero and $3D$, and $3D$ and $6D$. The maximum resolution of the instrument is less than one micron.

The instrument consists of a typical microdensitometer combined with an isophotometer. The isophotometer attachment consists of a specimen table drive, aperture adjustments, a recording table attachment, a recording pen assembly and a programmer (Fig. 2.25).

The isophotometer attachment to the original specimen table of the microdensitometer provides automatic parallel scanning according to a preselected programme. A stepping motor assembly advances the specimen table along a predetermined distance between successive parallel scans. The range available in the instrument is between 1·25 and 775 microns per step. In this manner it is possible to preselect the density of scans and to change it according to the requirements of the mapping programme during the operation of the instrument.

Two orthogonal independently variable field apertures were provided in the path of the beam emerging from the negative before impinging on the photomultiplier. The vertical aperture controls the resolution of the instrument, while the horizontal aperture controls the sensitivity of the measurement. Both apertures are adjusted via calibrated knobs so that the light intensity reaching the photomultiplier can be accurately and repeatedly set at the most suitable values for a particular investigation of a photographic negative.

The table of the conventional type of microdensitometer is replaced by the isodensitracing recording table, while an isophote pen assembly replaces the original microdensitometer recording pen in order to provide automatic two-dimensional plotting. The isophote pen drive assembly contains a stepping motor at one end of the lead-screw, which advances the pen assembly in steps per pulse.

A programmer automates the basic instrument and allows selection of scan spacing, horizontal magnification and recording speed. The instrument can be programmed to deliver from one to thirty-one pulses independently to the gearbox of the specimen table and the recording pen in order to provide a variety of distances between scans on the specimen and between pen lines on the record. The steps per stroke can be adjusted to give an asymmetric tracing of the negative, by programming a disproportionate step width of the pen advance as compared with step width for the specimen table and the preselected magnification ratio. A microswitch automatically shuts off the instrument when the isophote pen has traversed the width of the recording paper.

While a conventional microdensitometer scans along a single

line and yields a graph of optical density versus displacement, the isodensitracer scans along successive lines using the so-called *dropped-line technique* common in contour map making from elevation information in stereo pairs of aerial photographs (Fig. 2.26).

Interchangeable encoder commutators in conjunction with grey wedge attenuators control the write-out mode of the pen. The recording pen changes its mode of writing whenever the density

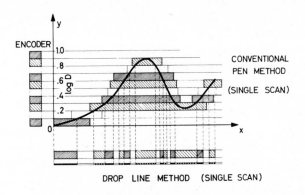

FIG. 2.26. The dropped line technique for making an isodensitogram.

changes by a discrete density increment preselected by the instrument. A line-dot-blank pattern is followed by the writing pen as the density decreases by preselected discrete amounts. For a density increase the writing sequence is reversed. The use of three recording symbols makes it obvious whether the density is increasing or decreasing as the recording mode changes. Density increments can be varied from $0{\cdot}005D$ to $0{\cdot}10D$ to provide accurate plotting of negatives of all densities. Since the writing mode of the recording pen in the isodensitracer changes at discrete density increments a single scan utilizing the drop-line technique is divided according to equal ordinate increments, which represent density variations.

The adjustable vertical aperture of the isodensitracer which controls the resolution of the instrument yields a final method of filtering high spatial frequencies of the image of moiré pattern. This can be achieved by averaging the light intensity transmitted through the negative transparency in the photoreading device. By sufficiently opening the vertical scanning slit of the instrument the contribution of higher harmonics to the overall intensity of the pattern is considerably reduced. By selecting a slit-width ten times the pitch of the gratings neither the individual rulings of the gratings appear in the scanned copy nor the influence of high harmonics of the signal is any more significant and the isodensitogram presents a pattern showing a sinusoidal variation.

For the accurate copying of the moiré pattern and the easy and faithful retrieval of the information by the isodensitracer scanning it is necessary to apply all the adjustments of the image-forming device, as well as those of the photoreading device used for the retrieval of the signal, which were previously mentioned, in order to eliminate the greatest part of high-frequency harmonics of the carrier signal and to transform it to its simplest form which is the sinusoidal curve. This elimination of the short wavelength phase variations of the signal does not influence the shape of the modulation of the carrier signal which alone contains the useful information of the moiré pattern from which the displacement components of the strain field can be extracted because this modulation is generally a long wavelength variation when compared to all wavelengths of the fundamental and the higher harmonics of the carrier.

While short wavelength can by these means be eliminated to a large extent, long wavelength phase variations as well as amplitude variations, due to uneven illumination combined with variations of the transmittance or reflectance of the specimen, do not interfere in the plotting of the two-dimensional map by the photoreading device since in an isodensitogram only relative fractional orders are traced between extrema. Therefore, neither the variations of absolute values of densities at these extrema due to background intensity changes, nor the changes in intensity amplitude of the signal influence the positions of fractional orders when they are traced by the instrument. This is a main advantage of the technique.

Another advantage is that the method readily yields the whole topographic map of the two-dimensional displacement field, as opposed to similar techniques[35] which necessitate several scannings along various traverses in order to yield a detailed picture of the displacement field along preselected points. Having a whole field picture of the displacement distribution it is possible to eliminate errors due to local disturbances which create regional unevennesses in the distribution of the same symbol areas in the isodensitogram. By smoothing out these areas a major part of these local errors is eliminated. Moreover, the tracing of an isodensitogram is rapid and totally automated. The internal inherent errors of the instrument are completely eliminated. A further advantage of the method is its versatility. Areas of a displacement field of particular importance (neighbourhoods of a crack tip or of a stress raiser) may be magnified and show a detailed picture of the displacement field for further minute study.

If all correction steps are taken into account in the image-forming device the intensity I_x of the image of the signal of the moiré pattern at the back of the camera normalized to its maximum fluctuation is sinusoidally modulated into the form

$$I_x = k + \sin \frac{2\pi x}{\lambda} \qquad (2.101)$$

where I_x is the intensity at a distance x from the start of the scan and λ is wavelength of the intensity variation. As the intensity cannot have a negative value for any value of x, the slowly varying quantity k, which expresses the background intensity, cannot be smaller than unity. The moiré signal in this form is projected on to the photographic emulsion, then, on development, this negative will have a transmission characteristic in the form

$$(I_x)^{-1} = \frac{1}{k + \sin \dfrac{2\pi x}{\lambda}}. \qquad (2.102)$$

[35] Sciammarella, C. A., op. cit., p. 52.

If this recorded image is exposed to a uniform light source and a transparent print is obtained, this print will have a transmission characteristic similar to the signal and it is expressed by eqn. (2.100). This double exposure of the copy of the signal is necessary in order to obtain a transparency in which the moiré fringes are varying sinusoidally. In this manner the extraction of the information from the sinusoidal carrier wave is much easier than in any other type of carrier wave. However, if k is much larger than the maximum amplitude of the sinusoidal wave (say five to six times this amplitude), expression $(I_x)^{-1}$ rapidly tends to eqn. (2.100) and this double exposure process can be avoided. Figure 2.27 shows the positions of twenty points along the abscissas which divide the ordinates of a half-wave of various $(I_x)^{-1}$-curves into equal segments. These points are compared with the corresponding points of a sinusoidal curve. For $k = 6$ it can be readily derived that the maximum deviation between the abscissas of corresponding points of the $(I_x)^{-1}$-curve and the sinusoidal curve is of the order of 5 per cent.

By using the positive transparency of the moiré pattern for the scanning process of the instrument it is possible to obtain an identical topographic map of the moiré pattern by the continuous scanning of the isodensitracer. But, if in this stage some higher harmonics persist to slightly modify the carrier system there is an adjustment of the instrument which allows a further elimination of these higher spatial harmonics of the carrier. This can be done by adjusting the vertical aperture of the instrument, as was mentioned previously.

Since the writing mode of the recording pen in the isodensitracer changes whenever the density of the photographic negative changes by a preselected amount on a discrete density increment, a single scan in the isodensitracer utilizing the drop-line method is divided according to equal increments of the ordinates, which represent density changes.

In order to determine which fraction of a moiré fringe corresponds to a certain area of the isodensitracer map, which belongs to the same recording symbol of the instrument, it is necessary to establish the areas of the map under the same recording symbol,

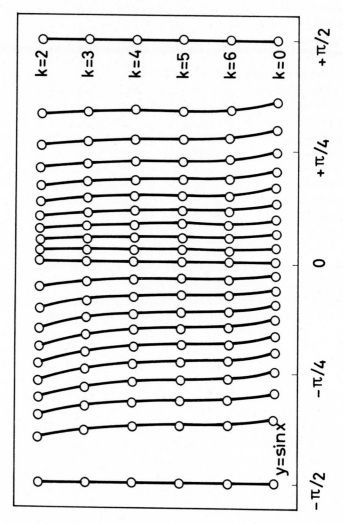

FIG. 2.27. Relative positions of abscissas of points dividing the ordinates of the half wave of the curves $\left(k + \sin \dfrac{2\pi x}{\lambda}\right)^{-1}$ into equal segments for different values of k.

which correspond to entire and half-order moiré fringes. Since these areas correspond to relative extrema in density, the sequence of symbols (lines-dots-blanks) in the isodensitogram must be reversed at these areas. Moreover, since the density variation of a moiré pattern follows a sinusoidal curve the plotted areas under the same recording symbol, which correspond to crests and troughs of the curve must be larger than the areas which correspond to intermediate densities, since the instrument is tracing contours of equal ordinate (density) increments. Then, these areas are characterized by a larger width than those of the intermediate areas.

However, since the instrument does not equally divide the density difference between successive density extrema, but indiscretely plots areas having a preselected density increment it is possible that the plotted areas corresponding to extrema belong to a fraction of the preselected density step.

In order to evaluate the density fractions of the extrema in the plotted isodensitogram the following calculations must be executed:

(i) The number n of the entire fractional density increments between successive extrema must be counted in order to

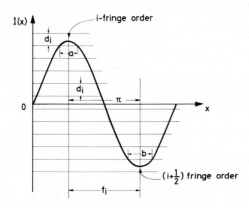

FIG. 2.28. Geometry of equal density division of a moiré fringe by the iso-densitracer.

evaluate the normalized standard density increment d_i (Fig. 2.28). For this purpose the density fraction corresponding to the fractional orders at the extrema is counted as an entire fractional order.

(ii) The widths a and b shown in Fig. 2.28 must be normalized by multiplying them by the factor $1/2f_i$, where f_i is the interfringe spacing between the (i) and $(i+i/2)$ moiré fringes.

(iii) From the normalized chords A and B $(A = b/2f_i)$ the corresponding values for the cosines are determined and compared to the values A_t and B_t of the abscissas of the first points in the cosine-curve divided to n-equispaced ordinates. The ratios A/A_t and B/B_t yielded the fractions of fractional orders at the extrema of the $(i, i+1/2)$-fringe. The values A_t and B_t are tabulated in the first and last columns of

n	$\sum_{i=1}^{n} \Delta x_i$												n
2	.250		.015	.015	.016	.016	.018	.019	.020	.023	.030	.070	21
4	.167	.083		.017	.017	.018	.018	.021	.021	.025	.032	.073	19
6	.134	.062	.054		.019	.020	.020	.020	.024	.026	.033	.078	17
8	.114	.050	.043	.043		.021	.022	.024	.024	.029	.036	.083	15
10	.101	.044	.037	.035	.033		.024	.025	.029	.030	.040	.089	13
12	.091	.041	.033	.031	.028	.026		.029	.032	.035	.043	.097	11
14	.086	.037	.030	.027	.024	.024	.022		.036	.040	.048	.108	9
16	.080	.035	.027	.024	.024	.021	.021	.018		.048	.056	.124	7
18	.075	.033	.025	.022	.021	.019	.019	.018	.018		.072	.146	5
20	.072	.030	.024	.021	.019	.019	.017	.016	.016	.016		.196	3

TABLE 2.9.1. Abscissas of points dividing a cosine-curve into n equal ordinate intervals normalized to the half-wave length of the curve.

Table 2.9.1 for n varying between unity and twenty-one. The two fractions of fractional orders added to the $(n-2)$ entire intermediate fractional orders yielded the exact number of fringes n_{ex} in the $(i, i+1/2)$ interval.

The exact position of each intermediate fractional order in the displacement-distance curve can be determined by evaluating the ratio of the average width w_k measured in the isodensitogram and the corresponding width for the same order of fringe w_{ok} given in Table 2.9.1. If there is no variation of displacement in the $(i, i+1/2)$ interval, the abscissa of the kth order fractional moiré fringe is given by kf_i/n_{ex}. In this case the average width of each intermediate fractional order w_k is approximately equal to the corresponding width w_{ok} given in the table for the closer entire number of fringes to n_{ex}. If the ratio w_k/w_{ok} is different than unity there is a variation of displacement in the $(i, i+1/2)$ interval and the position of the kth fractional fringe is given by $(k-1+w_k/w_{ok})f_i/n_{ex}$.

In the case where the intensity variation of the moiré pattern looks like a saw-tooth curve truncated either at the crests or at the troughs of its ordinates due to a different value of transmittance than 50 per cent, it is reasonable to accept that the density variation in the isodensitogram is also truncated at the corresponding crests or troughs and a number of fractional moiré fringes is eliminated at these areas. This number of fringes can be derived from the ratios A/A_t and B/B_t.

As an illustration the method was applied to two problems of linear elasticity. The first problem was of a thin strip subjected to uniaxial tension. A perforated strip subjected to longitudinal uniaxial tension constituted the second problem. The perforation was taken symmetric and its diameter was equal to the semi-width of the plate. The simple tension problem was chosen to show the linearity of data and to check the accuracy of the displacement-distance curve derived from the fractional moiré fringes. The second example was deliberately selected since analytic and experimental solutions of this problem exist, which accurately yield the stress and strain distribution throughout the field. Therefore, this problem can be used to evaluate the accuracy of the method.

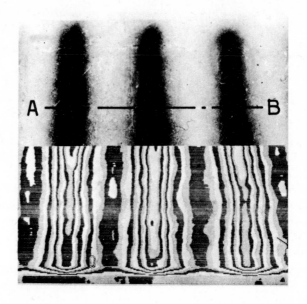

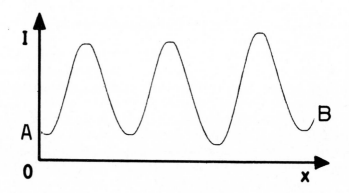

FIG. 2.29. Moiré pattern, isodensitogram and a microdensitometer scan along the traverse *AB* of the pattern in an epoxy resin plain specimen subjected to uniaxial tension.

The specimens were made from cold-setting pure epoxy polymer sheets of a thickness $t = 5\cdot1$ mm on which crossed gratings of a line pitch $p_s = 0\cdot050$ mm were cemented.

The widths of the specimens were for the plain tension specimen $2b = 40$ mm and for the perforated strip $2b = 50$ mm. The perforation in the strip had a diameter which was half the width of the specimen. Reference line gratings of identical line pitch were used to form evenly illuminated pictures of the specimens before loading. Small loads were applied in both problems which formed sparse moiré fringes. The positive transparencies of both loaded patterns were scanned in the isodensitracer. Figures 2.29 and 2.30 show the moiré patterns and the respective isodensitograms for the plain tension specimen and the perforated strip respectively. In the same figures microdensitometer scans along traverses of the specimen are presented.

Figure 2.31 shows the displacement curve for the plain tension specimen as it is derived from the isodensitogram. The straight line was satisfactorily compared with the theoretical displacement curve derived from the applied load on the specimen, its geometry and the mechanical properties of the material.

Figure 2.32 presents the displacement curve for the perforated strip along the boundaries as it has been derived from the isodensitogram and compared with the displacement values given by the theoretical solution.[36]

The comparison of the results of the two problems gave a high accuracy for the experimental values derived by the isodensitracer, the larger discrepancies detected over the whole displacement field were of the order of 3 to 5 per cent.

In order to estimate the sensitivity of results by using the isodensitracing process a plain tension specimen was used with a grating of a line-frequency of 40 lines/mm. The reference grating has a line-frequency of $39\cdot96$ lines/mm and therefore an initial disparity between the two gratings existed which gave an $\varepsilon = 0\cdot0010$ mm/mm[37] when the two gratings were superposed with their

[36] Howland, R. C. J., *Phil. Trans. Roy. Soc., London,* **A 220,** 838 (1930).
[37] Theocaris, P. S., *Proc. Amer. Soc. Test. Mat.* **61,** 838 (1961).

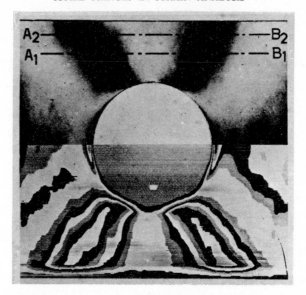

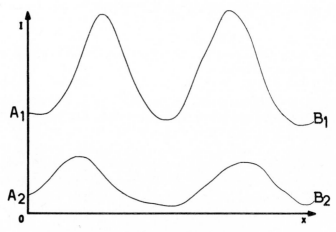

FIG. 2.30. Moiré pattern, isodensitogram and microdensitometer scans along the traverses A_1B_1 and A_2B_2 of the pattern in a perforated strip subjected to uniaxial tension.

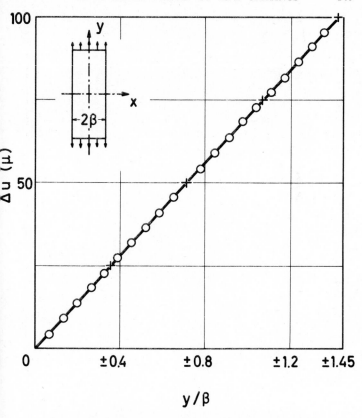

FIG. 2.31. Displacement curve along the longitudinal axis of the plain tension specimen derived from the isodensitogram (O experimental points, + theoretical points).

principal directions in coincidence. The unloaded moiré pattern showed an interfringe spacing f_o given by

$$f_o = \frac{p(1+\varepsilon)}{\varepsilon} = 25 \text{ mm}$$

The loaded moiré pattern was scanned by an isodensitracer and the combined merits of the differential method supplemented by

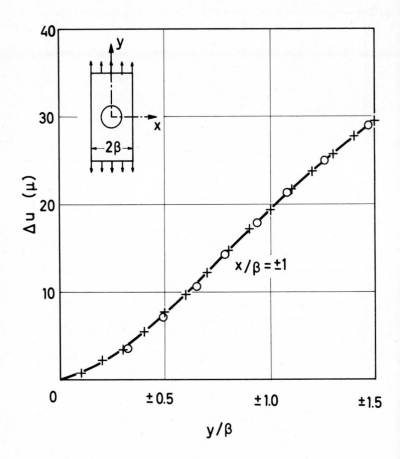

FIG. 2.32. Displacement curve along the boundaries of a perforated strip subjected in simple tension as it has been derived from the isodensitogram (O experimental points, + theoretical points).

isodensitracing were estimated. The 1/20th of a moiré fringe could easily be determined by the isodensitracing scan. For an applied load creating a loaded interfringe spacing which is reduced by the 1/20th of f_o the corresponding total strain e is given by

$$e = \frac{2 \cdot 5 \times 10^{-2} \, \text{mm}}{23 \cdot 75} = 1 \cdot 052 \times 10^{-3} \, \text{mm/mm}$$

The strain e_d due to the deformation field is given by

$$e_d = (e - \varepsilon) = 52 \times 10^{-6} \, \text{mm/mm}$$

Thus, the linear differential moiré method combined with the isodensitracing method can measure strains of the order of 50 μmm/mm or 50 μin./in. on a gauge length of 1 in. This sensitivity is very high and satisfactory for strain measurements.

The isodensitracing method, which may be used to obtain fractional moiré fringes, proved to be of real value in retrieving the necessary information from a moiré pattern.

The advantage of the method is that the fractional order moiré map traced in an isodensitogram yields a complete detailed view of the two-dimensional displacement field which is independent of variations of absolute values of densities in the master moiré pattern presenting a coarse view of the displacement field. Moreover, changes in the intensity amplitudes between fringes do not influence the positions of fractional orders mapped in the isodensitogram.

The tracing of the detailed moiré map is rapid and versatile. The instrument can be changed from a coarse to a detailed tracing when the instrument is under operation.

The real value of the method lies in cases of displacement fields where the detailed picture of the displacement distribution in a very limited area of the field (crack-tip, discontinuity) between successive moiré fringes is of great importance. In such cases, while all other methods fail to yield reliable results, this method can give a detailed picture of the displacement distribution in this area.

CHAPTER 3

Moiré Patterns formed by Circular, Radial and Zone Gratings

3.1. Moiré patterns formed by the superposition of equispaced concentric circular gratings

The theoretical and experimental investigation of the properties of moiré fringes was primarily confined to moiré fringes formed by the interference of two-line gratings. This was because almost all applications of moiré techniques were limited to the measurement of rectilinear displacements.

After the preliminary studies of Righi,[1] Ronchi,[2] and Raman and Datta,[3] Lehman and Wiemer[4] in 1953 considered the cases of moiré patterns formed by the superposition of radial and concentric equispaced gratings. They gave relationships between the pitches of the gratings and the interfringe spacings of moiré patterns by expressing the equations of the families of curves in indicial form. Similar studies were recently presented by Nishijima and Oster,[5] Oster, Wasserman and Zwerling,[6] and by Kostak and Popp.[7] Pirard[8] also studied the moiré patterns formed by the superposition of any combination of two gratings consisting of

---

[1] Righi, A., *Nuovo Cim.* **21**, 203 (1887); **22**, 10 (1888).

[2] Ronchi, V., La prova dei sistemi ottici, *Attual. Scient.*, No. 37 (N. Zanichelli, Bologna, 1925), Ch. 9.

[3] Raman, C. V., and Datta, S. K., *Trans. Opt. Soc. London* **27**, 51 (1926).

[4] Lehman, R., and Wiemer, A., *Feingeräte-Technik* **2** (5) 199 (1953).

[5] Nishijima, Y., and Oster, G., *J. Opt. Soc. Am.* **54** (1) 1 (1964).

[6] Oster, G., Wasserman, M., and Zwerling, C., *J. Opt. Soc. Am.* **54** (2) 169 (1964).

[7] Kostak, B., and Popp, K., *Strain* **2** (2) 5 (1966).

[8] Pirard, A., *Analyse des Contraintes*, *Mém. GAMAC* **5** (2) 1 (1960).

straight lines, radial lines, equispaced concentric circles and other types of gratings. The interfringe spacings and the other characteristic quantities of the moiré patterns formed by these gratings were given by parametric relationships.

Theocaris and Kuo[9] developed a theory for the moiré patterns formed by circular gratings by using differential relationships between the families of curves and the moiré patterns when concentric circular gratings of equal or different pitch were relatively displaced, as well as when the specimen grating was submitted to a deformation. Since circular gratings are insensitive to rigid-body rotations as well as to shear, it was shown that these are suitable for the measurement of displacement in directions normal to the rulings. They were used to simultaneously yield both displacement components in uniform strain fields by a unique measurement. It was also shown that this type of grating may form an excellent continuous moiré rosette.[10,11]

The phenomenon of interference of circular and line gratings and its application to strain analysis was studied by Theocaris and Kuo.[12] Orthogonal and equiangular moiré rosettes were introduced for the measurement of strain and these were compared to the continuous moiré rosette.

Let two amplitude gratings, consisting of equispaced concentric circles be face to face superposed with their centres at a distance $2c$ apart, the pitch of the specimen grating be equal to p, while the pitch of the reference grating be equal to $p(1 + \lambda)$, where λ is an infinitesimal quantity positive or negative. Both families of circles are indexed by k and l running from zero to plus or minus infinity. A system of Cartesian coordinates is referred to both gratings with the x-axis passing through the centres of the circles and its origin O at the mid-distance between the centres.

The equations of the families of circles in the specimen and reference gratings referred to in the above-mentioned coordinate

9 Theocaris, P. S., and Kuo, H. H., *Z. Angewdte. Math. Phys.* **17** (1) 336 (1966).

10 Theocaris, P. S., *Exp. Mech.* **5**, 105 (1965).

11 Theocaris, P. S., *J. Polymer Sci.* **A 3** (6) 2619 (1965).

12 Theocaris, P. S., and Kuo, H. H., *Exp. Mech.* **5** (8) 267 (1965).

system are given by

$$(x-c)^2 + y^2 = k^2 p^2, \\ (x+c)^2 + y^2 = l^2 p^2 (1+\lambda)^2. \Big\} \tag{3.1}$$

The indicial equation of the moiré pattern formed by the super-position of the two gratings is given by relation (1.1), i.e.:

$$k \pm l = m. \tag{3.2}$$

Eliminating k and l from eqns. (3.1) it is deduced that

$$\{[(x+c)^2 + y^2] + [(x-c)^2 + y^2](1+\lambda)^2 - m^2 p^2 (1+\lambda)^2\}^2 \\ = 4(1+\lambda)^2 [(x+c)^2 + y^2] \cdot [(x-c)^2 + y^2]. \tag{3.3}$$

The moiré fringes formed by a large displacement of SG relative to RG are shown in Figure 3.1 for the case where the pitch of RG is larger than the pitch of SG.

In the case of small relative displacement between the two gratings, relation (3.3) can be simplified by taking the square roots of both sides of the equation, rearranging the terms and introducing a new coordinate system for which $X = (x-c)$ and $Y = y$.

Thus it is deduced that

$$mp(1+\lambda) = (1+\lambda)(X^2 + Y^2)^{\frac{1}{2}} - [(X-2c)^2 + Y^2]^{\frac{1}{2}}. \tag{3.4}$$

In polar coordinates, R and Θ having the same origin as the new orthogonal system, relation (3.4) becomes

$$mp(1+\lambda) = (1+\lambda)R - R\left[1 - \frac{4c}{R} \cos \Theta\right]^{\frac{1}{2}}. \tag{3.4a}$$

Equations (3.4) and (3.4a) hold for all values of the displacement $2c$ between the two gratings. For small values of the displacement $2c$, the square root of the right-hand side of the above relation may be approximated, and thus the relation yields

$$mp(1+\lambda) = \lambda R + 2c \cos \Theta. \tag{3.5}$$

Equation (3.5) represents a family of cardioids which are shown in Figure 3.2.

In order to define the commutation moiré boundary (see pp. 12

and 17) the expression for $\psi(x,y)$, given by eqn. (1.10), must be obtained in this particular case:

$$\psi(x, y) = \frac{x^2 + y^2 - c^2}{p^2(1+\lambda)\{[(x+c)^2 + y^2][(x-c)^2 + y^2]\}^{\frac{1}{2}}}. \qquad (3.6)$$

FIG. 3.1. Moiré pattern formed by a large relative displacement of two equispaced circular gratings of different pitch.

The commutation moiré boundary is defined by $\psi(x,y) \equiv 0$. From relation (3.6) it results that the equation of the commutation moiré boundary is

$$x^2 + y^2 = c^2. \qquad (3.7)$$

Equation (3.7) shows that the circumference of a circle of

diameter $2c$ is the commutation moiré boundary. Inside this circle the additive moiré pattern is effective, while outside the circle the subtractive moiré pattern is effective (Fig. 3.3).

FIG. 3.2. Moiré pattern formed by an infinitesimal relative displacement of two equispaced circular gratings of different pitch (moiré fringes are a family of cardioids).

In the case where the gratings are identical ($\lambda = 0$), eqn. (3.3) becomes

$$\frac{x^2}{p^2 m^2} + \frac{y^2}{p^2 m^2 - 4c^2} = \frac{1}{4}. \tag{3.8}$$

This relation holds for all values of the displacement $2c$ between the two gratings. Inside the commutation moiré boundary this relation represents a family of ellipses, which remains unaltered, in the case when $p^2m^2 - 4c^2 > 0$. Outside the commutation moiré boundary $p^2m^2 - 4c^2 < 0$ and the family of moiré fringes represents a family of hyperbolas with the centres of gratings as foci. Figure 3.3 shows a schematic representation of the two families of curves

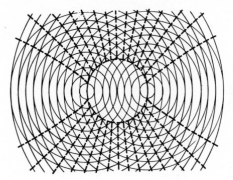

FIG. 3.3. Schematic representation of the formation of moiré fringes by the interference of two circular gratings of equal pitch. The commutation boundary coincides with the circle passing through the centres of the two families of circles. Inside this boundary moiré fringes constitute a family of ellipses, while outside the boundary they represent a family of hyperbolas.

while Fig. 3.4 is a photographic record of the moiré pattern formed when the relative displacement $2c$ between the two gratings is large.

For $\lambda = 0$, that is when the pitches of the two gratings are equal, eqns. (3.4a) and (3.5) reduce to

$$mp = R\left[1 - \left(1 - \frac{4c}{R}\cos\Theta\right)^{\frac{1}{2}}\right] \tag{3.9}$$

and

$$mp = 2c\cos\Theta. \tag{3.10}$$

While eqn. (3.9) is valid for all values of c, eqn. (3.10) holds only for small displacements c. Equation (3.10) represents a bundle of straight lines passing through the origin as shown in Fig. 3.5.

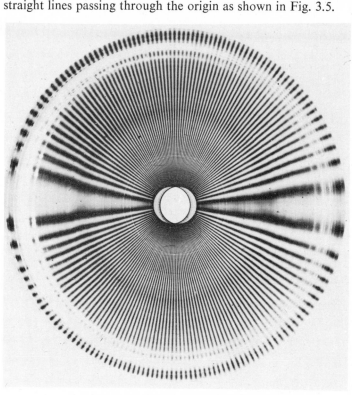

FIG. 3.4. Moiré pattern formed by a large relative displacement of two equispaced circular gratings of equal pitch (moiré fringes are hyperbolas with the centres of the gratings as foci).

In the limiting case where the distance between centres is zero, eqn. (3.3) reduces to

$$x^2 + y^2 = \frac{m^2 p^2 (1+\lambda)^2}{\lambda^2}. \qquad (3.11)$$

The moiré pattern is a family of concentric circles with radii $mp(1+\lambda)/\lambda$, where p is the pitch of the specimen grating, $p(1+\lambda)$

the pitch of the reference grating, and m a positive integer (Fig. 3.6). If the two families of concentric circles have the same pitch ($\lambda = 0$)

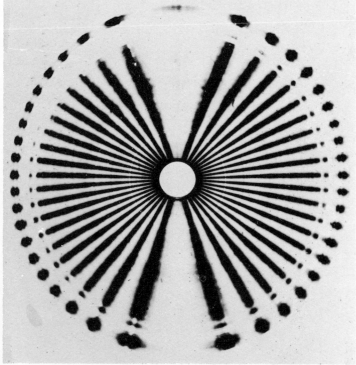

FIG. 3.5. Moiré pattern formed by an infinitesimal relative displacement of two concentric gratings of equal pitch (moiré fringes are straight lines passing through the middle of the centre distance of the two gratings).

the moiré circles are formed at infinity as can be deduced from eqn. (3.11).

3.2. Moiré patterns formed by combinations of circular, radial and line gratings

In the case where a grating consisting of concentric circles is superposed on to a parallel line grating it is assumed that the origin of the coordinate system coincides with the centre of circles. The equations of the two families of lines are

$$x^2 + y^2 = k^2 p^2, \\ y = lp(1 + \lambda). \tag{3.12}$$

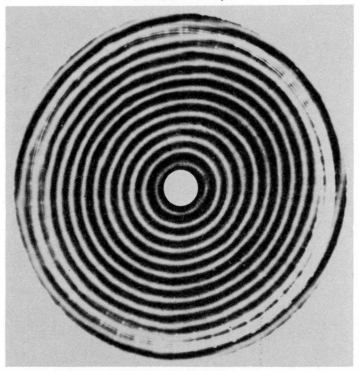

FIG. 3.6. Moiré pattern formed by the superposition of two circular gratings of different pitch when the intercentre distance of the gratings is equal to zero (moiré fringes are concentric circles.

The indicial equation of the moiré pattern formed by the superposition of the two gratings is found by eliminating the parameters

k and l between eqns. (3.12) and (1.1). The equation of moiré fringes is given by

$$(1+\lambda)^2 x^2 + \lambda(2+\lambda)y^2 \pm 2mp\,(1+\lambda)y - p^2(1+\lambda)^2 m^2 = 0. \quad (3.13)$$

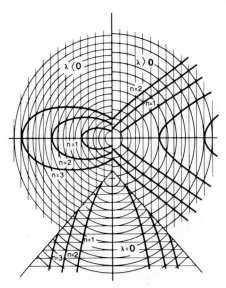

FIG. 3.7. Schematic representation of the formation of moiré fringes by the superposition of a line and a circular grating of different pitch (moiré fringes are hyperbolas, parabolas or ellipses for $\lambda > 0$, $\lambda = 0$ and $\lambda < 0$).

The plus sign corresponds to the additive moiré pattern, while the minus sign to the subtractive moiré pattern. It is easy to show that the x-axis is the commutation moiré boundary. The subtractive moiré pattern is valid for positive values of y, while the additive moiré pattern is valid for negative values of y. The moiré fringes of the additive pattern are symmetric to the corresponding fringes of the subtractive moiré pattern.

For λ positive the moiré pattern forms a family of ellipses, while,

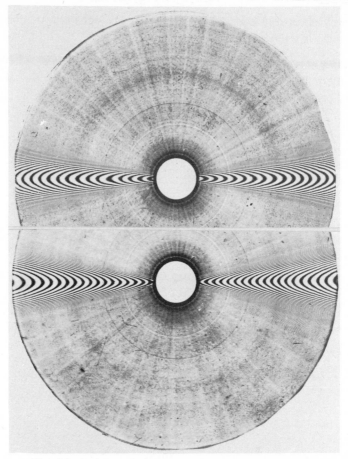

FIG. 3.8. Moiré patterns formed by the superposition of a circular grating and two line gratings with pitches larger (ellipses) or smaller (hyperbolas) than the pitch of the circular grating.

for λ negative, the moiré pattern is a family of hyperbolas. For $\lambda = 0$ the moiré pattern represents a family of parabolas. These results can be readily deduced from eqn. (3.12) by introducing the appropriate value for λ. Figure 3.7 gives a schematic representation of the three cases of moiré families, that is for $\lambda < 0$, $\lambda > 0$ and $\lambda = 0$, while Fig. 3.8 shows the moiré patterns formed by the superposition of circular and line gratings with $\lambda \gtrless 0$.

If a radial grating and a line grating are superposed, the moiré pattern formed by the two gratings may be found by eliminating the parameters k and l from relations

$$\left.\begin{array}{c} y = kp, \\ \text{arc tan } y/x = l\alpha, \end{array}\right\} \tag{3.14}$$

and relation (1.1).

The equation of moiré fringes is given by

$$\alpha y \pm p \,.\, \text{arc tan } y/x = mp\alpha. \tag{3.15}$$

In order to define the commutation moiré boundary the quantity ψ must be evaluated. This is given by

$$\psi = \frac{x}{p\alpha(x^2 + y^2)}, \tag{3.16}$$

which becomes zero for $x = 0$. Therefore the commutation moiré boundary is the y-axis. For positive values of x the subtractive moiré pattern is effective [the minus sign in relation (3.15)], while, for negative values of x, the additive moiré pattern is effective. Figure 3.9 shows the additive moiré pattern formed by the superposition of a line and a radial grating.

3.3. Moiré patterns formed by radial gratings

While the rosettes made either by pairs of circular gratings or by circular gratings and line gratings may be convenient for the measurement of uniform strain fields, in the cases of variable strain fields the information derived by using only circular gratings is not sufficient for the complete determination of the strain components. In order to complement the information deduced from circular gratings it is necessary to use another type of grating appropriate to yield data in polar coordinates as the data derived from concentric gratings. A convenient type of grating for this purpose is the radial grating. Radial gratings are also convenient to form moiré gauges and strain rosettes. Theocaris[13] showed the advantages of such a gauge over the linear moiré gauges for the measurement of the normal and shear components in a variable strain field.

For simplicity let two families of equiangular radial lines form two gratings and let the distance between the centres of the two bundles of lines be equal to $2c$ (Fig. 3.10). Also let the small angle

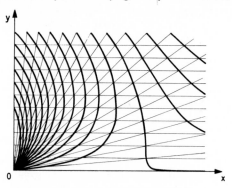

Fig. 3.9. Schematic representation of the formation of additive moiré pattern formed by the superposition of a line and a radial grating.

between two successive lines of each bundle be α and a system of Cartesian coordinates referred to both gratings having its x-axis passing through the centres of both bundles with origin O at the mid-distance between the centres.

The equations of the two families of bundles are given by

$$\left.\begin{array}{l} \arctan\left(\dfrac{y}{x+c}\right) = k\alpha, \\[4mm] \arctan\left(\dfrac{y}{x-c}\right) = l\alpha. \end{array}\right\} \qquad (3.17)$$

[13] Theocaris, P. S., *J. Sci. Instrum.* (*J. Phys.* E), Ser. 2, **1** (6) (1968).

The indicial equation for the moiré pattern given by eqn. (3.2) may be separated into two equations corresponding to the additive and subtractive moiré patterns respectively:

$$(k + l) = m \qquad (3.18)$$

and

$$(k - l) = n \qquad (3.19)$$

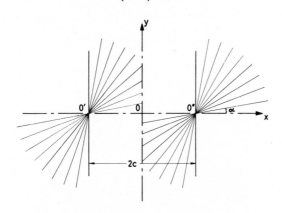

Fig. 3.10. Geometric representation of two bundles of equiangular straight lines.

Eliminating k and l between eqns. (3.17) and (3.18) the indicial equation for the moiré pattern is

$$x^2 - y^2 - 2xy \cot m\alpha = c^2. \qquad (3.20)$$

Similarly, for the case of the subtractive moiré pattern eliminating k and l between eqns. (3.17) and (3.19) the indicial equation for the moiré pattern is

$$x^2 + y^2 + 2cy \cot n\alpha = c^2. \qquad (3.21)$$

In order to define the position of the commutation moiré boundary the expression $\psi(x, y)$ given by eqn. (1.10) must be evaluated. For the case of radial bundles ψ takes the value

$$\psi(x, y) = \frac{x^2 + y^2 - c^2}{\alpha^2 [(x+c)^2 + y^2][(x-c)^2 + y^2]}, \tag{3.22}$$

which becomes zero for

$$x^2 + y^2 = c^2. \tag{3.23}$$

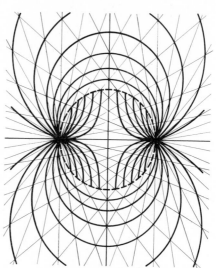

FIG. 3.11. Schematic representation of the formation of two families of moiré fringes by two bundles of equiangular straight lines mutually displaced by a distance $2c$.

Therefore, the commutation moiré boundary is a circle of radius c with centre the origin of the coordinate system. Its circumference passes through the centres of the two bundles. Inside the circle the additive moiré pattern is predominating and is composed of families of hyperbolas with their centres at origin O (eqn. 3.20). Outside the commutation moiré circle the moiré pattern is composed of a family of circles having their centres on the y-axis and their circumference passing through the centres of the two bundles (eqn. 3.21). Figure 3.11 gives a schematic representation of the formation of moiré patterns. Figure 3.12a shows the family of hyperbolas formed by the superposition of two radial gratings (additive moiré pattern), while Fig. 3.12b shows a moiré pattern of a family of circles, the circumferences of which pass through two points not shown in the figure (subtractive moiré pattern).

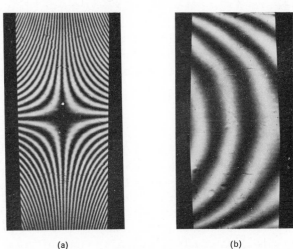

(a)   (b)

FIG. 3.12. Moiré patterns formed by the superposition of two sectors of radial gratings and consisting either of (a) families of equilateral hyperbolas with their asymptotes, or (b) of families of circles.

3.4. Moiré patterns formed by zone gratings and equispaced line gratings

A zone grating is a concentric circular grating consisting of rings the radii of which vary proportionally with the square roots of integers. In such a grating the areas of the transparent and opaque rings are equal. If the intercentre distance between two zone gratings is equal to $2c$, the equation of the families of these figures with respect to a system of Cartesian coordinates which has its origin at mid-distance between centres and its x-axis passing through the centres of the circles, is given by

$$(x-c)^2 + y^2 = ka^2, \Big\}$$
$$(x+c)^2 + y^2 = lb^2. \Big\} \tag{3.24}$$

It is assumed in these relations that the area of each additional transparent or opaque ring is taken equal to the respective areas πa^2 and πb^2 of the central circles of the two gratings. Of interest is the case where $a = b$. In this case, the moiré pattern formed by the superposition of two relatively displaced zone gratings by a distance $2c$ is found by eliminating the parameters k and l between eqns. (3.24) and (1.1). The equation of moiré patterns is expressed as

$$[(x+c)^2 + y^2] \pm [(x-c)^2 + y^2] = ma^2. \tag{3.25}$$

In order to define the commutation moiré boundary between the additive and the subtractive moiré patterns [eqn. (3.25) with the plus or minus sign] the quantity $\psi(x,y)$ must be evaluated. In the case of two zone gratings ψ becomes

$$\psi(x, y) \equiv x^2 + y^2 - c^2 = 0, \tag{3.26}$$

which is zero at the circumference of a circle of radius c which passes through the centres of the concentric circles of the two families. Outside this circle the subtractive moiré pattern is effective, while inside it the additive moiré pattern is predominant.

The indicial equation for the subtractive moiré pattern is found from eqn. (3.25) by taking the case with the minus sign. Then

$$x = \frac{ma^2}{4c}. \tag{3.27}$$

This equation expresses a family of straight fringes parallel to y-axis whose interfringe spacing is $f = \dfrac{a^2}{4c}$.

The indicial equation for the additive moiré pattern is found from eqn. (3.25) by taking the case with the plus sign. Then

$$x^2 + y^2 = \frac{ma^2 - 2c^2}{2}. \tag{3.28}$$

Equation (3.28) represents a family of concentric zone circles having their centre at the origin of the coordinates and radii depending on the distance $2c$ between the centres of the two families of initial circles as well as on the radius a of the internal circle of each grating. The difference between the zone circle family

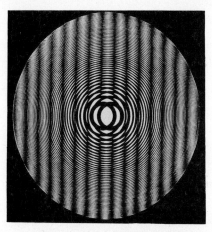

FIG. 3.13. Moiré pattern formed by two zone gratings of inner-most circles of equal radius a displaced by a small distance (subtractive type of moiré fringes consisting of parallel straight lines).

of the additive moiré pattern and the initial zone gratings is that the area of their innermost circles is different.

While the subtractive moiré pattern consists of parallel equi-spaced fringes for close centre-to-centre distances $2c$, the additive moiré pattern manifests itself for centre-to-centre distances $2c$ much larger than the radius a of the internal circle of each zone grating. The area of existence of the additive moiré pattern is restricted to the circle of radius c, with centre the origin of the coordinate system.

Figure 3.13 shows the straight equidistant fringe pattern when two gratings are relatively displaced by a small amount, while Fig. 3.14 shows the additive moiré pattern appearing at the centre-to-centre area of two zone gratings overlapped so that $2c > a$.

The case of a zone grating superposed on to a line grating yields

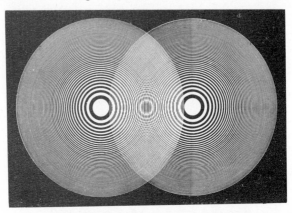

FIG. 3.14. Moiré pattern formed by two zone gratings of innermost circles of equal radius a displaced by a distance $2c$ larger than the radius a (additive type of moiré fringes consisting of a similar zone grating of different internal circle radius).

interesting moiré patterns. In this case the equations of the two families of lines are

$$\left. \begin{array}{l} x^2 + y^2 = ka^2, \\ x = lp. \end{array} \right\} \tag{3.29}$$

In order to justify the existence of multiple moiré patterns the indicial eqn. (1.1) is written in this case as

$$(k \pm nl) = m, \tag{3.30}$$

where n takes values of any positive or negative integer.

Eliminating k and l from eqns. (3.29) and (3.30) yields the equation of the moiré pattern:

$$\left(x \pm \frac{na^2}{2p}\right)^2 + y^2 = ma^2 + \left(\frac{na^2}{2p}\right)^2. \qquad (3.31)$$

The commutation moiré boundary is defined by evaluating the quantity $\psi(x,y)$ given by eqn. (1.10). The quantity $\psi(x,y)$ for the case of a zone grating superposed with a line grating is

$$\psi(x, y) = 2x. \qquad (3.32)$$

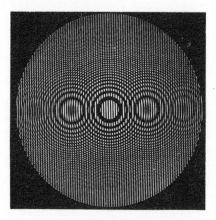

FIG. 3.15. Moiré pattern formed by the superposition of a zone and a line grating.

The quantity $\psi(x,y)$ becomes zero for $x = 0$, i.e. along the y-axis. The subtractive moiré pattern is effective for positive values of x, while the additive moiré pattern is effective for negative values of x. The multiple moiré pattern is then symmetric about y-axis.

Equation (3.31) represents, for a given value of n, merely the equation for a single zone grating displaced by a distance $(na^2/2p)$ from the original origin of the coordinate system, which coincided with the centre of the zone grating. Moreover, the phase of the original grating is shifted in the moiré pattern by a distance equal to $(na^2/2p)^2$. Hence, for a given zone grating and for a given value of n a phase shift of a half period takes place for a value of pitch of the line grating equal to $p_e = na^2/\sqrt{2}$. Figure 3.15 shows the moiré pattern formed by the superposition of a line grating on a zone grating.

3.5. Determination of the strain components from moiré fringes

For the measurement of the strain components in a variable strain field it is necessary to dispose two complementary types of gratings. If it is convenient to express the variable strain field in polar coordinates, it is necessary to use a circular grating complemented by a radial grating. In this case an entire radial grating is needed, the rulings of which must cover an angle of 2π. The radial grating is superposed to the circular grating so that their centres coincide. Such a pair of gratings may yield with high accuracy the polar components of strain. In addition, since these gratings are insensitive to angular displacements and rigid-body rotations of the specimen, they are also convenient for determining the strain ellipse and the directions of the principal strain axes of the field.

Consider the case of a circular and a radial grating related to a system of polar coordinates r, ϑ for which the r-direction for $\vartheta = 0$ coincides with the x-direction of the respective Cartesian coordinate system. The only component of strain which can be completely evaluated from a circular grating is the ε_{rr} radial strain. The circular grating also yields the γ_{rr} shear term and the $\varepsilon_{\vartheta r}$ term of the tangential strain. These components for identical reference and specimen gratings before deformation are given by relations (2.12) or (2.32), which for the case of circular gratings become

$$\varepsilon_{rr} = \frac{\partial u_r}{\partial r} = \frac{p_c}{f_{cn}}, \quad \varepsilon_{\vartheta r} = \frac{u_r}{r},$$

and
$$\gamma_{rr} = \frac{\partial u_r}{r\partial\vartheta} = \frac{p_c}{f_{ct}}, \qquad \qquad (3.33)$$

where p_c is the pitch of the reference circular grating, f_{cn} is the interfringe spacing of the two adjacent moiré fringes bounding the point of measurement and measured along the radius connecting this point with the origin, and f_{ct} is the interfringe spacing of the same moiré fringes measured on the arc of the circular ruling passing through the point. If the distance between adjacent moiré fringes is small, this may be measured on the tangent to the arc at the point. If this distance is large, measurement of the chords of successive arcs on the ruling yields a satisfactory approximation.

For the measurement of the displacement u_r the components of the ε_{rr} strain along the radius passing through the point must be integrated and the displacement of a reference point at the free boundary of the specimen must be known. Graphical integration of the ε_{rr} strains between the reference point and the point of measurement along the radius r passing through this point yields the displacement u_r which, divided by the distance r, yields the $\varepsilon_{\vartheta r}$ component of strain.

The moiré pattern formed by two radial gratings with identical angles subtended by two successive rulings of the gratings yields the tangential component of strain $\varepsilon_{\vartheta\vartheta}$ and the other term $\gamma_{\vartheta\vartheta}$ of the shear component.

These terms are given by

$$\left.\begin{aligned} \varepsilon_{\vartheta\vartheta} &= \frac{p_r}{f_{rt}}, \\ \gamma_{\vartheta\vartheta} &= \frac{p_r}{f_{rn}} - \frac{v_\vartheta}{r}, \end{aligned}\right\} \tag{3.34}$$

where p_r is the pitch of the reference radial grating at the point of measurement expressed by $p_r = r\alpha$, f_{rn} is the interfringe spacing of the two adjacent moiré fringes bounding the point and measured along the radius connecting the point to the origin, and f_{rt} is the interfringe spacing of the same moiré fringes measured on the arc of the circle passing through this point. For the measurement of the displacement v_ϑ the components of the $\varepsilon_{\vartheta\vartheta}$ strain along the circumference passing through the point must be integrated between a reference point of this circumference, the displacement of which is known, and the point of measurement. Graphical integration of the $\varepsilon_{\vartheta\vartheta}$ strains between these points yields the displacement v_{ϑ} which, divided by the distance r, yields the v_{ϑ}/r component of the shear strain.

Besides the above-mentioned use of circular and radial gratings to yield the polar components of strain in a variable field, the use of these gratings as moiré gauges seems to be very suitable for measurements of uniform or slowly varying strain fields.

Crossed-line and circular gratings can be used if the strain field has sufficiently large uniform sections. If a crossed-line grating is used as a specimen grating, which covers the area of measurement, and a crossed-line grating is used as a reference grating it is possible to measure by this gauge the Cartesian components of the normal and shear strains and from them to evaluate the principal strain components and directions.

If, instead of a crossed-line grating, a circular grating of a sufficiently large diameter is cemented on to the surface of the specimen so that it covers the gauge area and a second circular grating of different pitch is used as reference, the set of the two superposed gratings forms the so-called *continuous moiré rosette*. This gauge has the advantage over the line moiré gauge to be insensitive to rigid-body rotations and angular displacements. It directly yields the principal strain directions, which coincide with the largest and shortest axes of the deformed gauge. If the strain components are to be evaluated along these directions, they are found directly without any further calculations. Figure 3.16 shows a continuous moiré rosette in its undeformed and deformed states. The rosette measures the deformation of a plain tensile specimen made of an epoxy resin and it simultaneously yields not only the longitudinal elongation of the specimen but its lateral contraction due to Poisson's ratio effect. The principal strain directions may be easily traced by defining the largest and shortest diameters of the moiré curves.

It is sometimes advantageous for the measurement of the components of strains in a uniform strain field to use a circular grating as a specimen grating and a series of line gratings as reference gratings. By properly orienting the reference gratings the radial components of strain can be measured along the principal directions of the reference grating. If the reference gratings are oriented to form a 90° or 60° or 45° crossed grating they can form together with the circular grating a rectangular, equiangular or T-Δ strain rosette.

It is worth while mentioning here an effect created with a 60°

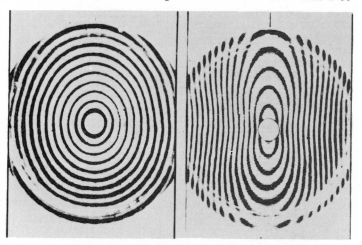

Fig. 3.16. Continuous moiré rosette in its undeformed and deformed states formed by two circular gratings of different pitch superposed centre to centre.

crossed-line grating. Superposing two line gratings of equal frequency with a circular grating of different frequency, three moiré patterns are formed instead of two. The moiré patterns are identical patterns angularly displaced by an angle of 60°. The third moiré pattern is formed by the interference of the moiré pattern formed from the two line gratings interacting independently with the circular grating. The interference of the two line gratings at 60° formed a moiré pattern of the same frequency as the two initial gratings. Since the interfringe spacing of this moiré pattern is very sensitive to the angle of the two initial gratings, slight angular displacement of the initial gratings changes the shape of moiré fringes and eventually causes their transition from a family of hyperbolas to a family of ellipses. Since this transition is very

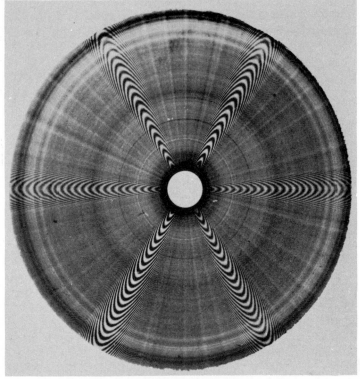

FIG. 3.17. Equiangular moiré rosette formed by the interference of a circular and two line gratings forming an angle of 60°.

unstable, it yields a sensitive means of measuring angles. A similar phenomenon is created with a 45° moiré rosette. Figure 3.17 shows the moiré pattern formed by an equiangular moiré rosette.

Versatile moiré gauges may be also formed by radial gratings. In this case parts of sectors of radial gratings are usually used, taken at large radius R, in order to give rather uniformly spaced rulings. These sectors are superposed in opposite directions along the centre-to-centre line. Therefore additive moiré patterns are generally formed (Fig. 3.12a).

If two parts of sectors are superposed in the same direction they form parts of circumferences belonging to the circles of the subtractive moiré pattern. But, since the distance $2c$ is generally small compared to the radii of the circles, these parts of circumference resemble a family of parallel straight lines parallel also to the intercentre line of the original gratings. Therefore these do not present any additional interest in strain analysis to the fringes formed by line gratings (Fig. 3.12b).

An alternative possibility to the case of using two radial gratings in opposite directions, which is more convenient in practice, is to superpose a radial grating of average line pitch p_a and a line grating of the same pitch. In this case the equation of moiré fringes is given by relation (3.15).

Consider a square section of side l of a radial grating of large mean radius which covers the gauge area of the specimen assumed to be uniformly strained. This radial grating constitutes the specimen grating and it is convenient to have two such gratings orthogonally crossing each other in order to be able to define all strain components. A similar section of the same grating is used as a reference grating. This is successively superposed to the specimen grating in two orthogonal directions and in opposite directions to the respective radii of the specimen gratings, so that each pair of gratings, during each superposition, forms a moiré pattern of families of equilateral hyperbolas and their asymptotes (Fig. 3.18). If p_a is the average pitch at mid-length of the square gauge, the pitch of each grating is variable and it is equal to

$$p_x = p_a\left(1 \pm \frac{x}{R}\right), \qquad (3.35)$$

where x is the distance from the asymptote, which is taken as the axis of the gauge, and R is the radius of the grating at mid-length of the gauge.

The maximum values for p_x are

$$p_{max} = p_a\left(1 \pm \frac{l}{2R}\right). \tag{3.36}$$

The variation of pitch along the gauge length constitutes thé main advantage of the radial moiré gauge. Thus a radial moire

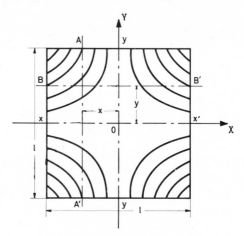

FIG. 3.18. Geometry of a radial grating
moiré gauge.

gauge corresponds to an infinite number of line gratings of variable pitch. The interfringe spacing f_y at a section AA' is given by

$$f_y = \frac{p_a\left(1 \pm \dfrac{x}{R}\right)}{x/R} \approx \frac{p_a R}{x}, \tag{3.37}$$

while the number n_y of fringes at a section AA', which is at a distance x from the asymptote, is given by

$$n_y \approx \frac{l}{p_a R}|x|. \tag{3.38}$$

The maximum number of fringes at the boundaries of the gauge is

$$n_{max} \approx \frac{l^2}{2p_a R}.$$ (3.39)

Indeed, from Fig. 3.12a it is clear that the moiré fringes in all boundaries of the gauge and along any traverse parallel to the asymptotes are equispaced.

The position of the xx' asymptote corresponds to the straight line along which the pitches of the two gratings are equal, while the position of the yy' asymptote corresponds to the points along which one ruling from one grating coincides with another ruling of the other grating.

If the reference grating has undergone an infinitesimal rigid-body translation with respect to the specimen grating it is easy to show that this translation does not modify the form of the fringes. These are only shifted by the same infinitesimal amount along the same direction.

If the reference grating has undergone an angular displacement ϑ about origin O the moiré pattern will be shifted in a direction parallel to the xx' asymptote by a translation equal to (ϑR). Thus, the angular displacement of the moiré pattern due to an angular displacement of RG is multiplied by the radius R of the grating. Simultaneously the pattern will be displaced by an angle ϑ due to rigid-body rotation, but this quantity is infinitesimal as compared to the multiplied displacement ϑR, and is therefore negligible.

Let the specimen grating undergo a perceptibly constant deformation over the gauge area of a radial square gauge of side l. Let the components of the strain tensor be ε_x, ε_y and γ_{xy} and the direction of ε_y coincide with the yy' asymptote of the moiré pattern. The ε_y strain evidently has no effect on the moiré pattern since the displacement y, which engenders this strain, is strictly parallel to the rulings forming the asymptote yy' and approximately parallel to all other rulings of the grating and, therefore, it results only in a shifting of the moiré pattern by a quantity equal to the displacement without amplification.

The ε_x strain results in a translation of the moiré pattern by displacing the xx' asymptote corresponding to the line of equality of pitches of the two gratings; this translation is multiplied by the radius R of the gauge and it is given by

$$\eta_1 = R\varepsilon_x. \tag{3.40}$$

The effect of the γ_{xx} shear strain term is an angular displacement of the grating equal to γ_{xx} radians. The moiré pattern undergoes a translation along the xx' asymptote, which is equal to the γ_{xx} term of the applied shear strain γ_{xy} multiplied by the radius R of the gauge. Then

$$\xi_1 = R\gamma_{xx}. \tag{3.41}$$

By measuring the corresponding translations of the asymptotes of the gauge along the two principal axes of the gauge it is possible to evaluate the two components of strains ε_x and γ_{xx}.

In order to evaluate the remaining components of strains ε_y and γ_{yy} it is necessary to dispose a second radial specimen grating oriented at right angles to the first one. Similarly, by the relative displacements of its asymptotes, this grating yields

$$\eta_2 = R\gamma_{yy}, \tag{3.42a}$$

$$\xi_2 = R\varepsilon_y. \tag{3.42b}$$

Therefore, the shear component of strain is given by

$$\gamma_{xy} = R^{-1}(\xi_1 + \eta_2). \tag{3.43}$$

If the specimen grating (SG) is a crossed radial square grating and the reference grating (RG) a simple radial grating identical to SG, this grating must be successively superposed to each of the radial gratings of SG to yield the four strain components evaluated from relations (3.40)–(3.43). The radius R of the grating, which defines the amplification of the measured components of strain, can be determined from either of relations (3.41) or (3.42a) by counting the number of fringes n_y or n_{max}, either at any section AA' of the gauge, at a distance x from the asymptote, or at the boundaries of the square gauge. The average pitch p_a of the grating is necessary to be known in advance.

While the measurement of the normal components of strains is accurate and independent of any relative angular displacement between RG and SG, the measurement of the two partial shear components γ_{xx} and γ_{yy} is subject to error due to a misalignment of the superposed gratings. However, this error is considerably reduced in radial gratings because angular displacements do not significantly influence the positions of the asymptotes. An accurate definition of the asymptotes before and after loading may almost completely eliminate the influence of angular misalignment to the measurement of the shear components. Therefore the pair of asymptotes plays the role of axes of reference whose orientation remains unaltered during deformation of the gauge.

In the case of a line grating superposed on to a radial grating, relations (3.35)–(3.39) are valid and, moreover, relations (3.37)–(3.39) are exact. The only difference between the moiré patterns formed by two radial gratings and, on the other hand, by a radial and a line grating is that, in the latter case, the moiré fringes are no longer equilateral hyperbolas but curves of the second degree, whilst the asymptote xx', corresponding to the locus along which the pitches of the two gratings are equal, is no longer a straight-line but a second-degree curve. Since the radial gratings are parts of sectors of radial gratings taken from a large radius R so that their rulings may be approximated to parallel rulings, the xx' asymptote may also be considered as a straight line.

The use of a line grating as a reference grating has the advantage over the case of two radial gratings superposed that the relative position of RG and SG does not influence the moiré pattern and a parallel shifting of RG on the xx' asymptote does not change the moiré pattern.

Radial moiré gauges consist of square sections of radial gratings (either single or double, orthogonally crossed, in which case an orthogonal rosette is formed yielding all components of strain at a point) photographically reproduced on stripping films. The photographic reproduction of gratings with an average density between 1000 and 2500 lines per inch is possible and is described in detail in Chapter 11. The procedure of reproduction of the specimen grating on the surface of the specimen follows one of the methods described in the same chapter. For illuminating the moiré pattern formed by *RG* superposed on *SG* two incandescent lamps should be used, placed about 2 feet from the specimen. Satisfactory moiré patterns may be photographed with this illumination and a camera placed some distance from the specimen without resorting to any lens system provided an oil layer is placed between the specimen and the reference gratings in order to avoid air gaps between the gratings and eliminate higher-order interferences.

With the reference grating properly positioned, the original moiré pattern with the specimen unloaded is photographed. Then the specimen is loaded and the loaded moiré pattern is photographed. The displacement of the moiré asymptote parallel to the applied displacement between loaded and original moiré patterns yields the value of the normal displacement and strain. The magnification *R* of the gauge can be evaluated from the original pattern by using one of relations (3.35)–(3.39). Moreover, the displacement of the asymptote may be accurately evaluated by counting the number of fringes at the gauge boundaries parallel to the asymptotes. After a deformation of the specimen grating the number of fringes at these boundaries will be different and the distance of the new position of the asymptote from the boundaries can be determined by using eqn. (3.39). A check of the accuracy of the values of the distances of the new position of the asymptote after loading is that their sum must equal to twice the distance of the initial position of the asymptote from the same boundaries.

In order that the gauge is capable of measuring the shear components of strain, the reference grating, reproduced on an Estar film, must be cemented with the specimen grating along the asymptote *yy'*. This asymptote is normal to the applied direct strain and is formed by the total coincidence of one ruling of the reference grating with a corresponding ruling of the specimen grating. Then the displacement of this asymptote, which is solely due to the shear component of strain, yields by the technique mentioned above the value of the component.

By repeating the procedure with a gauge oriented 90° to the first or by using a radial orthogonally crossed grating as the specimen grating, the two other components of strain can be evaluated and the entire strain-field at the point of the gauge determined. Figure 3.19a shows the original moiré pattern of a simple radial moiré gauge used to measure the deformation of an aluminium strip under simple longitudinal tension. Figure 3.19b shows the loaded moiré pattern after an application of a uniform strain field equal

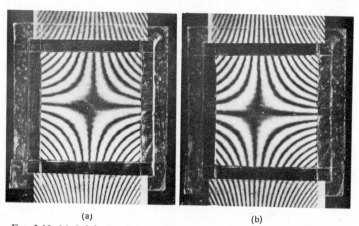

(a)   (b)

FIG. 3.19. (a) Original and (b) loaded moiré patterns obtained by a radial moiré gauge measuring the deformation of an aluminium strip subjected to pure tension.

to $\varepsilon_s = 0.00765$ strain. This strain was measured by an electric strain gauge attached to the specimen. The dimensions of the moiré gauge were $l = 40$ mm and the magnification $R = 1070$ mm. The average pitch of the gauge was $p_a = 0.050$ mm, and the value of the strain determined from the moiré gauge was equal to $\varepsilon_m = 0.0075$ strain.

The gauge was calibrated to different values of normal strains and the overall discrepancy between values of electric strain-gauge readings and moiré-gauge readings was less than 3 per cent.

With the above gauge, which presents a number of 15 fringes at its boundaries, when unloaded, and by taking into consideration that an increase of a tenth of a fringe along each boundary can be easily detected, the sensitivity of the gauge can be evaluated by using relations (3.39) and (3.40) and is

$$\varepsilon_{min} = 2 \cdot 5 \times 10^{-4} \text{ strain.}$$

If a grating of larger magnification radius is used and, moreover, a diffraction grating with much smaller pitch p_a, the sensitivity of the gauge can be considerably increased.

While the use of radial and circular gratings for the measurement of the components of a strain field seems to be very promising, the use of zone gratings in strain analysis seems rather impracticable.

Consider two zone gratings superposed so that they form a subtractive moiré pattern which consists of a family of parallel straight lines normal to the two gratings, the minimum pitch p_{min} is expressed by

$$p_{min} = \frac{a}{2\sqrt{n}}, \tag{3.44}$$

where n is the order of the extreme circle of the gratings. In addition,

$$R = a\sqrt{n},$$

then

$$p_{min} = \frac{a^2}{2R}. \tag{3.45}$$

For a shifting h of the one zone grating relatively to an identical zone grating the interfringe spacing f_h is given by

$$_h = \frac{a^2}{4h}. \tag{3.46}$$

Therefore, there will be $(2R):(a^2/4h) = (8hR)/a^2$ fringes along the diameter $2R$ of the grating.

For an infinitesimal translation dh of the extreme fringes of the grating the variation of interfringe spacing df_h is expressed as

$$df_h = -\frac{a^2}{4h^2} dh,$$

and for the $(8hR)/a^2$ fringes the total variation of the distance between extreme fringes is

$$-\frac{8hR}{a^2}\,\frac{a^2 dh}{4h^2} = -\frac{2R}{h}\,dh. \tag{3.47}$$

Thus the amplification of the zone grating due to infinitesimal shifting of the two gratings is equal to $(2R)/h$. This amplification is only valid for infinitesimal variations of h because the quantities measured are not linearly varying with respect to dh.

If the diameter $2R$ of the gratings is multiplied by an arbitrary factor by maintaining the condition of constancy of the minimum pitch p_{min} it may be easily deduced that the initial innermost area πa^2 is also multiplied by the same factor. If the initial interfringe spacing is kept constant, the initial value of h, which is proportional to a^2, is multiplied by the same factor. The amplification $(2R)/h$ is independent of R for a given p_{min} and f_h. Therefore the only advantage for a large radius R is the possibility of having a higher initial value for h and consequently a larger latitude of measurement.

It is worth while pointing out that:

(a) The variation of the distance between extreme fringes is not linear with respect to the translation dh.

(b) The latitude of measurements with this gauge is very limited.

Therefore the use of moiré patterns formed by zone gratings is only convenient for the measurement of infinitesimal rigid-body translations.

CHAPTER 4

Applications of Line Gratings to Two-Dimensional Strain Measurement

THE potentialities of the moiré method to two-dimensional strain analysis are demonstrated in this chapter by two examples. In both cases crossed gratings were used as specimen gratings together with a line grating as the reference grating. The reference grating was superposed twice with its principal directions coinciding with either of the principal directions of the line gratings forming the crossed specimen grating. Since the problems treated present a geometric and loading symmetry, errors due to misalignment of the reference gratings are minimized.

The first illustrative example is concerned with the strain distribution in the neighbourhood of the neck in a plain tensile sheet specimen.[1] In this type of problem, where the components of strain are large, the equal pitch moiré method was used. Since the test was executed at ambient temperature it was possible to place the reference grating in contact with the surface of the specimen.

The second example is concerned with the application of moiré method to the study of the elastoplastic deformation of thin polycrystalline sheet specimens for the determination of the true stress–strain curves of the materials and the curves of lateral contraction ratio versus strain or stress.[2,3] Here the linear differential moiré method was used since the magnitudes of strain in the elastoplastic domain were expected to be small. As, apart from the longitudinal

[1] Theocaris, P. S., and Marketos, E., *Proc. First Int. Conf. Fracture, Sendai, Japan* **2**, 1781 (1965).

[2] Theocaris, P. S., and Koroneos, E., *Phil. Mag.* **8** (95) 1871 (1963).

[3] Theocaris, P. S., and Koroneos, E., *Proc. Am. Soc. Test. Mater.* **64**, 747 (1964). See also Theocaris, P. S., *Exp. Mech.* **4** (8) 223 (1964).

strains, the transverse components are also needed for the evaluation of the lateral contraction ratio, a crossed grating was used as specimen grating. The reference line gratings, with reduced or enlarged pitches to different magnitude, were successively placed in contact with the specimen along the two principal directions. For the case of tests at high temperatures the specimen grating was imaged at unit magnification on the ground-glass screen of a camera, where the reference grating interfered with the image of the specimen grating to form the moiré pattern. Therefore the second illustration describes a problem where the linear differential method is applied with either the contact or the image interference techniques.

4.1. The equal-pitch moiré method applied to strain distribution of a necked tensile sheet specimen

The specimens tested were cut from a thin sheet of a low-carbon high-yield-strength alloy steel, under the commercial designation USS T1. This type of steel alloy was chosen because it presents a reduced work-hardening characteristic which considerably helps the formation of neck. Moreover, the fine-grained metallic sheets of this alloy used in the experiments did not present an appreciable anisotropy in several directions.

Oversized longitudinal samples were cut from larger and thicker sheets and were machined to a uniform thickness of 4 mm. The final width of the specimen at mid-length was 90 mm and the length was 700 mm. A small taper of an angle of 2° to the longitudinal axis was given in order to form a minimum section at the mid-length of the specimen. The two tapered parts were joined with an arc of circle of a radius of 270 mm. Hence the large fillet along each longitudinal boundary of the specimen assured a minimum section where the neck was forced to appear, without introducing appreciable stress concentration.

The last layer of thickness of 0·25 mm was removed by slow grinding with a well-lubricated porous wheel in order to avoid an increase in temperature and introduce surface stress during grinding. The longitudinal axis of the specimen was chosen such that it coincided with the rolling direction of the sheet. The surface of each specimen and the part corresponding to the gauge area was finally ground by a medium–fine grinding wheel and then polished with diamond paste. Thus the surface of the specimen was finished free from pits, scratches or other imperfections.

The specimen gratings used were orthogonally crossed gratings of a density of 500 lines per inch. On to this grating a reference line grating of initially equal pitch was successively superposed in the two principal directions. Since large plastic strains at the vicinity of the neck were to be measured, originally identical gratings gave dense patterns after loading.

The reproduction of the specimen grating on to the surface of the specimen was achieved by spraying the gauge area of the specimen with a Kodak Photo Resist (KPR) photosensitive thin coating and allowing the specimen to dry in a horizontal position. The coating was exposed to an arc light source whilst in contact with the negative of the specimen grating. The developing, dyeing and drying processes were then carried out. Etching was not carried out as it was judged unnecessary since the experiments were to be conducted at ambient temperature. As a reference grating a line grating of the same pitch was used reproduced on a photographic glass plate.

The 100-ton Amsler hydraulic testing machine was used, having a capacity considerably larger than that required for the particular experiments. Thus the inertia effects of the weighing mechanism, as well as of the elasticity of the different components of the testing machine on the shape of the stress–strain curves, were very small. Crosshead speeds were kept low and constant during the application of load, interrupted at constant load intervals for taking photographs after each additional load step.

The testing machine was fitted with axial loading shackles specially designed to assure alignment of the applied load by allowing free rotation in three mutually perpendicular directions in both shackles. Self-alignment was obtained through arrangements of thrust bearings and ball joints in each shackle. The alignment of the specimen in the grips of the testing machine was checked by loading each specimen up to the yield point and observing that the emerging moiré fringes from the one extremity of the gauge area run along parallel lines. This proved to be a very sensitive way of checking the alignment. In addition, the deformation speed was controlled by the speed of emergence of the moiré fringes.

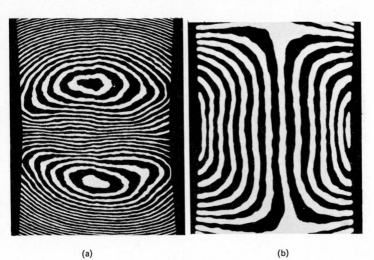

(a)   (b)

FIG. 4.1. Moiré patterns of the displacement fields along the axes of symmetry of a tensile specimen at a loading step antecedent to the maximum load.

Five incremental loading steps were chosen for each test. The two first steps corresponded to loads smaller than the maximum load, and the subsequent three steps were related to the evolution of the neck.

The direct superposition of the reference grating on to the deformed specimen grating gave moiré fringes which yielded the displacement components along the principal directions of the specimen. Figure 4.1a, b shows the displacement fields in both principal directions for the step of loading before the initiation of neck and Fig. 4.2a, b presents the moiré patterns at a loading step

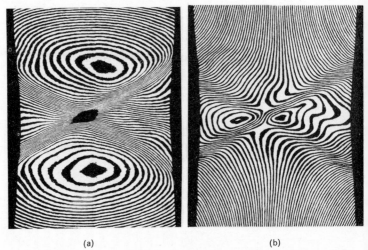

(a)   (b)

FIG. 4.2. Moiré patterns of the displacement fields along the axes of symmetry of a tensile specimen at a loading step forming a triple neck.

following the initiation of neck. Both figures cover the whole gauge area of the specimen.

For each loading step the strain distributions were evaluated along fourteen transverse and longitudinal sections. These are shown in Fig. 4.3. In cases where the strain distribution was similar

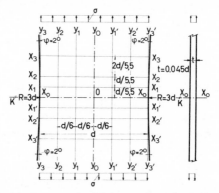

FIG. 4.3. Gauge area of the tapered tensile specimen.

in the two halves of the specimen the values of strains of the upper
half appear in the discussion of results. The displacement versus
distance curves along the lines of the coarse grid were traced by

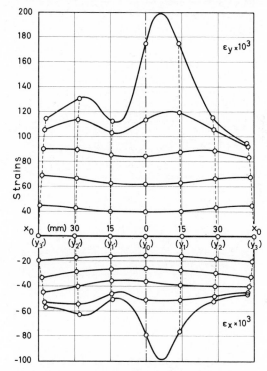

FIG. 4.4. Longitudinal ε_y and transverse ε_x
strain distribution along the transverse section
x_0–x_0 of the tensile specimen for the five loading steps.

following the technique described in section 2.2. The u- and
v-displacement curves along all the lines of the coarse grid were
graphically differentiated in order to yield the four components of
strain $\partial u/\partial x$, $\partial u/\partial y$, $\partial v/\partial x$ and $\partial v/\partial y$. Curves of all these derivatives
versus distances along both families of lines of the coarse screen

are traced and smoothed out. At each knot of the coarse screen all four direct and cross derivatives of displacement were evaluated

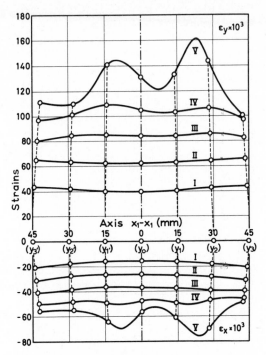

FIG. 4.5. Longitudinal ε_y and transverse ε_x strain distribution along the transverse section x_1–x_1 of the tensile specimen for the five loading steps.

from these curves and the components of strain were calculated by using either relations (2.20) or (2.35).

The longitudinal ε_y and the transverse ε_x strains are shown in Figs. 4.4–4.7. The strain distribution is presented for the five successive loading steps and along four sections parallel to the minimum section x_o–x_o of the specimens.

It is clear from these diagrams that the ε_x strain distribution is similar to the ε_y strain distribution. The curves of the ε_y and ε_x strains are convex downward at the middle of the x_o-x_o section for loads preceding the maximum load P_{max}. This distribution degener-

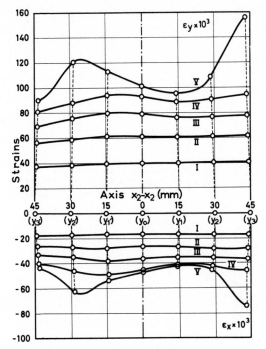

Fig. 4.6. Longitudinal ε_y and transverse ε_x strain distribution along the transverse section x_2-x_2 of the tensile specimen for the five loading steps.

ates progressively to the inverse distribution with withdrawal from the minimum section. The same conclusion may be deduced from Fig. 4.8, where the curves of equal ε_y strain distribution appear for the three first steps of loading.

For loads preceding or equal to P_{max} an increase in strains ε_y and ε_x at the vicinity of the minimum section and close to the free boundaries results in a strain distribution which appears to be

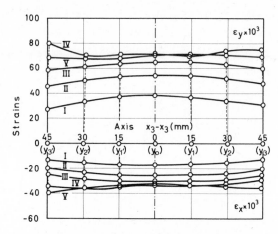

FIG. 4.7. Longitudinal ε_y and tranverse ε_x strain distribution along the transverse section x_3–x_3 of the tensile specimen for the five loading steps.

locally concave at the neutral axis. Local protrusions in the strain distribution appear at the vicinity of the boundaries, which move as moderate strain waves from the free boundaries to the interior, which increase in value with increasing load. The non-uniform

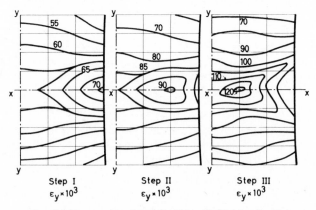

FIG. 4.8. Isoentatics of the ε_y strain distribution for the firs t three loading steps.

strain distribution across any transverse section becomes more severe for the maximum load P_{max} and increases further for the subsequent steps of loading. The peaks of the moving strain-protrusions and of the severe local raisings of Figs. 4.4–4.6 correspond to the final steps where visible necks were formed.

It may be concluded from the above analysis that the occurrence, which predicts the initiation of the necks and the subsequent fracture, appears long before the applied load reaches its maximum value.

Another important observation is that the neck first appears in places different from its final position and it moves progressively toward its final position with the evolution of the applied load.

Figure 4.8 shows the ε_y strain distribution during three characteristic load steps. The evolution of the phenomenon is followed during the interval when a longitudinal strain of the order of 70×10^{-3} mm/mm appears first and then disappears from the gauge area studied. It is worth while noting that the maxima of ε_y strains move continuously, while they increase with the applied load, from the boundaries to the interior. This movement is followed by the neighbouring strains in a V-shape with the apex coinciding with the position of the maximum deformation and oriented toward the longitudinal axis. The angle of the V-shape was of the order of $29°$. In this manner the apices of the ε_y strains, moving from the free boundaries to the interior from both sides of the longitudinal axis, merge at the vicinity of this axis forming a double neck.

Figure 4.9 presents the distribution of the ε_y strains during the last loading step. Three necks appear in this figure which, according to their size, are characterized as primary, secondary and tertiary neck. The strain distribution along the necks is highly non-uniform. The apparently non-symmetric shape of necks and the presence of a tertiary neck may be attributed either to a small eccentricity of the external load or to a non-symmetric and non-uniform strain distribution due to geometrical or strain-hardening factors or, finally, to a combination of all these factors.

The initiation of the visible necks with severe plastic deformations of the order of 130×10^{-3} mm/mm does not occur at their final position. The bottoms of the necks, where the maximum strains appear, move towards the interior like moderate waves, in the same manner as the previously described movements of the maxima of the instability strain concentrations. This phenomenon may be deduced from Fig. 4.9 where the isoentatics (lines of equal

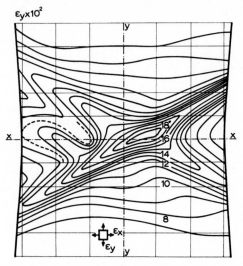

Fig. 4.9. Distribution of the ε_y strains during the last loading step where necking is already formated.

strain) appear denser at the edges of the neck lying at the interior of the specimen, which is the direction of movement of the neck, and sparser in the regions where the moving strain wave has already passed.

Thus, the movement of strain protrusions continues in the form of compression after the occurrence of the necks. The angle formed between the axes of necks with the transverse axis of symmetry of the specimen was equal to $\pm 29°$.

Figure 4.10 presents the evolution of ε_y strains along the lines $\alpha\beta$

and $\alpha'\beta'$, which define the bottom lines of the necks during the last step of loading. It is clear from these diagrams that there is no significant difference between the ε_y strains along the two bottom lines during the first two steps. On the contrary, there is a distinct difference between the strain distribution along these lines for step III. This difference becomes more severe in the last loading

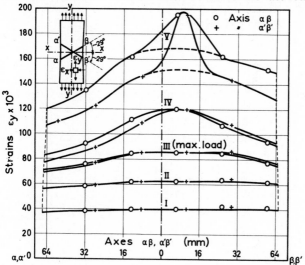

FIG. 4.10. Evolution of ε_y strain distribution along the bottom lines of the primary $\alpha\beta$ and secondary $\alpha'\beta'$ necks.

step. As it is expected at the crossing point of the two necks the ε_y strain takes equal values.

The coalescence of the protrusions moving from both sides was not accomplished until step IV. In the final step, when this coalescence has taken place, an abrupt increase in the curve of strain distribution occurs. The dotted lines in Fig. 4.10 show the shape of the curves of the ε_y strain distribution just before the coalescence of the two protrusions.

Then it may be concluded that the strain protrusion becomes remarkably intense at the area of coalescence of the two protrusions with the result that the specimen produces a visible neck and reaches closer to the point of rupture of continuity of the material than in any other part.

Figure 4.11 shows the shear strain distribution on the surface of

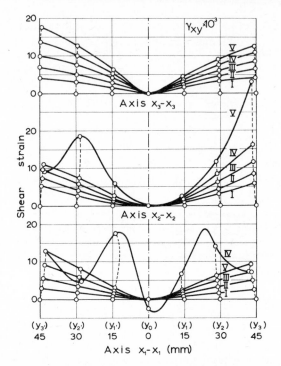

FIG. 4.11. Evolution of γ_{xy} strain distribution at transverse sections x_0-x_0, x_2-x_2 and x_3-x_3 of the tensile specimen for the five loading steps.

the specimen. Shear strains take small values and slightly increase during loading steps preceding the formation of the neck. The shear strain distribution is uniformly and smoothly changing. It becomes wavy and reaches larger values during the final step of the initiation of the neck.

The values of shear strains γ_{xy} shown in Fig. 4.11 on the lines

y_3, y_3', which are parallel to the longitudinal axis of the specimen, do not correspond to the values of shear strains at the tapered free boundaries of the specimen. In reality, these values correspond to the shear-strain components lying inside the parallel axes y_3, y_3' because the evaluation of displacement components in both sides of these axes and especially outside them is impossible due to the poor quality of reproduced specimen gratings at the vicinity of the boundaries.

However, there exist components of shear strain γ_{xy} parallel to the undeformed axes of symmetry of the specimen because of the tapered form of the specimen from its mid-length to its extremities (Fig. 4.3). Since the taper is increasing with increasing loading of the specimen, the values of shear components γ_{xy} are increasing from step to step of loading. This increase of taper is accelerated with the introduction of the neck due to the constriction of the specimen at the minimum section.

Moreover, the values of the shear strains γ_{xy} contain a term due to rigid-body rotation of the specimen during deformation of the specimen and especially at the vicinity of the necked areas where this rigid-body rotation is continuously changing. Since the reference grating was oriented to coincide with the specimen grating at a region with uniform strain far from the necked area, the continuous and unknown rigid-body rotation, created mainly at the necked regions during deformation, introduces an error on the calculated shear strain γ_{xy}. Since the rigid-body rotation at the neck is arbitrary and, moreover, is followed by kinks and abrupt discontinuities at the free boundaries, the amount of error introduced by the free-body rotation of the specimen cannot be estimated. In order to diminish the error due to the rigid-body rotation of the specimen, two moiré patterns were simultaneously taken at each step of loading with the reference grating matched either to the upper part of the specimen grating or to its lower part far from the necked area and the evaluation of shear strains was executed in parts from the corresponding moiré patterns. Therefore, the values of shear-strain components shown in Fig. 4.11 may be considered as sufficiently accurate.

4.2. The linear differential moiré method applied to the study of stress–strain curves of metals

As a second illustrative example the study of the elastoplastic deformation of plain tensile specimens made of polycrystalline sheet steel was chosen whilst determining the true stress–true strain curves of the materials, as well as their lateral contraction ratio curves versus true strain or stress. Two types of steel were tested. These were a low carbon steel under designation DIN St42 and a low carbon high-yield-strength alloy steel quenched and tempered, which had the commercial designation USS T1. This investigation was carried out[4] for ambient temperatures as well as for higher temperatures up to the glowing point of each type of steel.[5]

Since the investigation was concerned with the elastoplastic domain of deformation of metals, small increments of strain were to be accurately measured and, therefore, it was judged opportune to use the linear differential moiré method. It was necessary to measure simultaneously the longitudinal and the transverse components of strain at the elastic as well as the plastic domain of deformation. Since in these domains the distribution of strains is uniform all over the parallel section of the specimen, a large gauge area was utilized for the measurement of displacements. This fact, combined with the use of specimens of large dimensions, rendered the method very sensitive to strain variations.

Another phenomenon, which was studied by this method, and which is related with the transition state from elastic to fully plastic deformation of this type of steel, was the phenomenon of nucleation and propagation of plastic fronts and the study of the distribution of plastic strain in the enclaves, known as Lüders bands.

The specimens were cut as longitudinal samples from full hard-commercial-rimmed steel sheets and were machined to a uniform width. The flat sheets of 4–5 mm thickness had a gauge area

[4] Theocaris, P. S., and Koroneos, E., *Phil. Mag.* **8** (95) 1871 (1963).
[5] Theocaris, P. S., and Koroneos, E., *Proc. Am. Soc. Test. Mat.* **64**, 747 (1964).

270 mm long and 90 mm wide. The enlarged ends of the specimens were joined with the gauge area with large fillets so that the stress concentration at the fillet was kept small. The edges of the specimens were carefully rounded along the gauge area during final polishing. Great care was also exercised in properly aligning the specimens in the grips of the testing machine. All these precautions during the preparation of the specimens were judged indispensable for the reliability of results.

For the measurement of both components of strain an orthogonally crossed grating of a frequency of 500 lines per inch was printed on the gauge area of the reduced section of each specimen. The printing of the grating was made by a photoprinting process. The sides of the crossed screen were printed parallel to the principal axes of each specimen. After printing the grating a thin and uniform layer of a transparent protective varnish was sprayed on the surface of specimens.

Line gratings were reproduced on photographic plates with the rulings on the emulsion side of the plates coinciding exactly with the portion of the grating pattern photoprinted on the metallic surfaces of the specimens. This resulted in a coincidence of the existing pitch irregularities of the two gratings and consequently in a remarkable refinement of the moiré fringes formed by interference of the specimen crossed grating and the superposed reference line grating. The reference line gratings were reproduced with different pitch values. A number of reference plates were prepared with pitches gradually reduced from the standard pitch and the remainder with gradually enlarged pitches. Each of the reduced or enlarged reference plates when superposed on the specimen grating resulted in an initial moiré pattern with different values of interfringe spacing. The interfringe spacing became smaller as the difference in pitch between the two gratings was increased. The initial moiré pattern formed with a reduced reference grating, before loading of the specimen, could be considered as resulting from a fictitious elongation of the specimen grating. The amount of this elongation was such as to produce the existing difference in pitches between reference and specimen gratings. On the other hand, the initial moiré pattern formed with an enlarged reference grating could be considered as resulting from a fictitious reduction of the specimen grating. Therefore, reduced reference gratings are appropriate for the measurement of longitudinal extensions of the tension specimens, while the enlarged reference gratings are appropriate for the measurement of lateral contractions. Appropriate selection of the superposed reference grating during the various loading steps resulted in an interfringe spacing of the order of $\frac{1}{4}$–$\frac{3}{8}$ of an inch, yielding an optimum in accuracy of the evaluated strains.

As the type of testing machine used can affect materially the appearance of the stress–strain curves obtained, specially designed machines are needed for the recording of the abrupt fall from the upper yield point to the lower yield point level of mild steels. These machines must have little inertia so that small extensions of the specimen can produce large decrease in the stress, which must be recorded by a sensitive measuring device. Since the investigation was not concerned with the study of the special phenomenon of the production of the upper yield point instability, the tests were carried out in a 100-ton Amsler hydraulic testing machine. The testing machine had a capacity considerably greater than that of the specimen. Therefore the effects of inertia of the weighing mechanism, as well as the elasticity of the different parts of the testing machines on the shape of the stress–strain curves obtained, were very small.

The load was applied to the specimens via axial loading shackles, ensuring alignment of the applied load by allowing free rotation in three mutually perpendicular directions in both shackles. Before commencing a test, the axiality of loading was checked. This was done by loading and subsequent unloading of the specimen in the elastic range to within about 75 per cent of the elastic limit and observing the appearance of the moiré fringes. Uniform spacing of the fringes was a sensitive indicator of the uniformity of the strain field. Since the gauge area was extended all over the region where the crossed grating was printed and which occupied the central part of the reduced section of the specimen, uniformity of the strain field in this area proved the axiality of loading. Crosshead speeds, which influence considerably the reproducibility of the results, were kept constant during each test and very low, interrupted at constant time intervals for taking photographs after each additional increment. The speed of deformation was controlled by the speed of appearance of new moiré fringes at one end of the grating printed on the central part of the reduced section of each specimen. This proved to be a very sensitive speed gauge.

The moiré patterns formed by superposition in both directions of the suitably reduced or enlarged reference gratings on the specimen grating were photographed before loading and after each loading step with a fine grain polystyrene-base film. Transparent positives of these photographs were prepared on the same scale as the specimen dimensions. The intensity of light and the exposure time were selected to yield slightly underexposed photographs showing the darkest parts of moiré fringes.

For the complete evaluation of the strain field on the gauge area a coarse screen of equidistant lines was traced over the specimen and the history of deformation of the body at the various loading steps was followed by determining the components of strain at these reference points. Plots of the displacement curves along the lines of the coarse grid were constructed for each step of loading. By using a series of reference gratings of slightly different pitches a fictitious initial deformation was introduced in the loaded moiré pattern which resulted in dense and well-defined moiré patterns, although the pattern due to the actual deformation of the specimen contained only few and sparse moiré fringes. The fictitious displacements introduced by the linear disparities in pitch of each reference and specimen grating was linearly proportional to the coordinates of each point of the strain field and the resulting strain was constant all over the field. These strains were algebraically subtracted from the total corresponding strains given by each moiré pattern and yielded the strain distribution only due to the deformation of the specimen. By using a series of reference gratings at each loading step a series of moiré patterns with various moiré fringe densities were formed at different areas in the strain field.

In this manner the number of points in the displacement curves was increased and their plotting became accurate all over the strain field. The evaluation of the strain components from the various displacement versus distance curves was made by graphical differentiation and followed the procedures described in §§ 2.2 and 2.4.

The photographs of moiré patterns before loading yield the values of initial interfringe f_i from which the values of the coefficients λ_i were calculated by using relation (2.12). The displacements were measured in a gauge length of 70 mm in both the unloaded and the loaded patterns and in both directions. The large gauge length used was allowed by the uniformity of the strain field and resulted in a considerable increase in the accuracy of the measured strains. Plots of the displacement curves along the two principal axes were constructed for each step of loading. These curves coincided, in the elastic regions, with straight lines of different slopes. Strains were determined by using relation (2.37) in which the derivatives $\partial u_i/\partial x_i$ were evaluated from the slopes of the displacement curves and λ_i were derived from the initial interfringe spacing f_i, corresponding to the reference grating of pitch p_i, which was used at each loading step.

The results obtained from the moiré patterns at the corresponding load increments are discussed below whilst demonstrating the advantages of the moiré techniques.

The elastic range of straining was extended up to a well-defined yield point and no flow was observed below this point. On applying a small increment of load above the yield stress of the material plastic enclaves or Lüders bands were nucleated at one or both of the extremities of the reduced section of the specimen near the fillets due to the small stress concentration factor. The Lüders fronts were visible and countable through the moiré fringes formed on the surface of the specimen. While the band fronts moved into the specimen from each fillet, occasionally Lüders bands were nucleated at imperfections along the flat specimens, producing additional band fronts that move apart, along the flat-sheet specimens. Usually, all Lüders bands appeared as soon as the upper yield stress was applied but occasionally Lüders bands continued to form after the upper yield point was reached. A possible explanation may be that the larger-than-average grains in the specimen are potential sites for local yielding at a lower stress resulting from a previously formed Lüders band.

The yielded regions rapidly encompassed the cross-section of the specimen and the stress abruptly dropped to the lower yield point, which remained almost constant during the growth of yielded areas. The Lüders band or bands spread and progressively covered larger areas of the specimen as the displacement between the load transmitting shackles increased. The band spread over the specimen because local stress concentrations at the edge of the band assisted the applied load to reach the higher yield stress in the undeformed material neighbouring the Lüders front. Each yielding lamella of the specimen flowed under the applied stress by creep and it underwent practically the whole of the deformation. Thus the strain rate at each lamella next to the edge of the band was increasing abruptly.

The stress applied to the specimen remained approximately constant during the spreading of Lüders bands and only when the whole specimen was covered with the plastic bands did the stress–strain curve show a strain hardening for further plastic deformation. The strain corresponding to the starting of the strain hardening of the material is called Lüders strain or yield point elongation. Figure 4.12 presents Lüders bands for the two types of steel tested in various stages of Lüders deformation. The moiré patterns taken in the Lüders region show a strongly non-uniform displacement field. Strains were derived in this region through graphical differentiation of the displacement curves. In the fully plastic region the displacement field again became uniform and this uniformity extended up to the necking point of the specimen. In this region moiré fringes rapidly became very dense and it was necessary to change reference gratings after a few steps of loading. The resulting values of strains were very accurate as the influence of eventual defects of the gratings and misorientation in this stage of loading were insignificant. Figure 4.13 presents longitudinal and

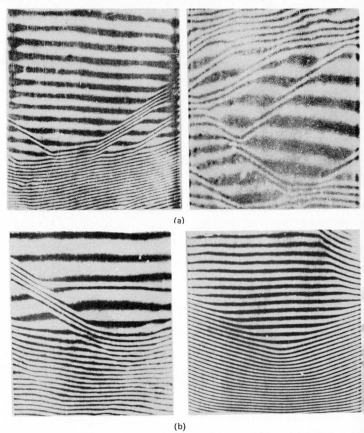

(a)

(b)

FIG. 4.12. Moiré patterns of Lüders bands at different stages of
evolution.

transverse moiré patterns in the plastic range of deformation of
two types of materials tested. The uniformity of the displacement
field is remarkable.

In all specimens the elastic strains were also determined by
means of SR4 electric strain gauges of type A8, disposed in longitudinal and transverse directions on the reverse side of the specimen and at the central region of the reduced section. The electric

strain gauges were employed until the gauges failed, usually at a longitudinal strain of about 0·3 per cent for all tests. The coincidence of the values of strains measured by the two methods in the

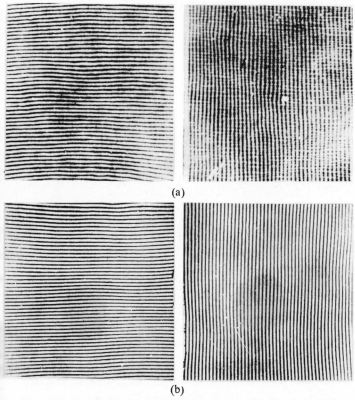

(a)

(b)

FIG. 4.13. Longitudinal and transverse moiré patterns in the plastic range of deformation of steel (a) DIN St42 and (b) USS T1.

elastic region is remarkable. Figures 4.14 and 4.15 show the conventional stress versus conventional strain, as well as the true stress–true strain curves for the two types of specimens tested.

Conventional or engineering stresses and strains are defined as the stresses and strains which are referred either to the initial section or to an initial gauge length before any application of external load on the specimen. True stresses and strains are referred to actual sections and gauge lengths respectively after the application of the various steps of loading.

In Figs. 4.14 and 4.15 the stress–strain diagrams were divided into two parts, i.e. the parts corresponding to the elastic and plastic regions of deformation of the specimen. The elastic longi-

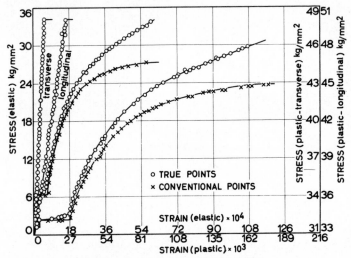

Fig. 4.14. Longitudinal and transverse conventional and true stress–strain curves for steel DIN St42.

tudinal and transverse strains are plotted with different and much larger scales than the corresponding quantities when the specimen has undergone large plastic deformation. This was done in order to plot the stress–strain diagrams with higher accuracy and to show the versatility of the moiré method for detecting elastic as well as plastic deformations. The elastic parts of the stress–strain curves are referred to the left-hand scales of the figures, while the plastic parts are related to the right-hand scales. In the elastic regions both conventional and true stresses and strains are coincident and, therefore, are shown with the same signs.

The load-deformation curves obtained for both types of steel tested were almost of the type with a small exhibited peak stress at the beginning of the Lüders region. The major portion of the Lüders strain occurred at almost the same stress. The important observation is that the stress corresponding to the major portion of the Lüders region is approximately the same for each type of

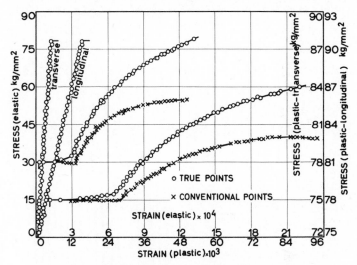

FIG. 4.15. Longitudinal and transverse conventional and true stress–strain curves for USS T1.

steel. When the whole gauge area has been covered by Lüders bands the stress–strain curves rose again according to the strain hardening of each material. Although no other irregularities in the strain distribution were observed during this region of load-ing the remnants of some markings of Lüders bands persisted and it was not until some higher stress was reached, correspond-ing eventually to the reappearance of the upper yield point stress, that these irregularities in the pattern disappeared (Fig. 4.16).

Conventional and true stress–strain curves deep in the plastic range far from the instability of the Lüders region presented the characteristic shapes of such curves influenced by the strain hardening of each material.

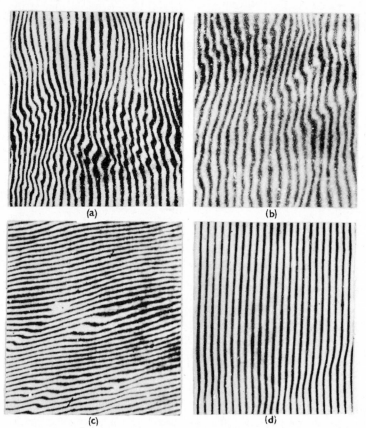

Fig. 4.16. After Lüders' irregularities in the displacement field of the two types of steel tested.

Very little indication of the presence of a pronounced upper yield point was disclosed in the tests. This is in agreement with previous observations according to which tests on flat-sheet specimens do not usually show pronounced upper yield points as do round bars. Although the load decreased somewhat with the propagation of Lüders bands, the stress remained constant to within an amount which was appreciably smaller than experimental reproducibility of the yield stress.

The local strain within a well-developed Lüders band was not uniform, being a little greater at the centre of the plastically

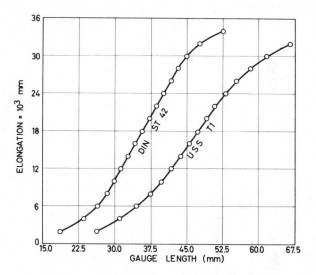

FIG. 4.17. Distribution of Lüders' elongation along the longitudinal axis of the two types of steel tested.

deformed zone and characteristically smaller at the band front. The distributions of Lüders' elongation along the specimens are given in Fig. 4.17 for an advanced stage of Lüders' evolution. The reduction in the strain observed in a Lüders band, together with the reduction in the velocity of the band-front motion, explains the phenomenon of the reduction in extension rate as the band moves along the specimen.

Therefore, it is proved by these tests that the decrease of the extension rate with time is a consequence of a decrease of the band-front velocity and the decrease of the strain associated with the band. When two different band fronts meet, no sharp decrease in the extension rate is observed. On the contrary, the junction and annihilation of Lüders fronts propagating in opposite directions resulted in a smoothing of the non-uniformly strained area and it did not remain an important permanent defect.

Measurements of the orientation of the planar fronts on the specimen surfaces yielded the following results. Always single-band fronts, first to appear, formed an angle with the longitudinal axis of the specimen, which was of the order of 50°. As the Lüders band was extending on the surface of the specimen and new fronts appeared behind the primary front, the angle of these fronts with the tensile axis increased constantly. Thus angles have been measured varying between 50° and 60°, which is the angle that the neck forms with the tensile axis.

Examination of the inclination of moiré fringes in the first and subsequent Lüders bands shows a definite distortion of fringes due to shear. The angle of inclination of fringes in the band was constant all over the area of a simple Lüders band and at an angle between 30° and 40° with the tensile axis. When the band front was complex consisting of two bands of different inclination to the axis, the angle of inclination of moiré fringes diminished from the boundary to some point at the interior of the specimen where the two bands were adjoined (Fig. 4.12a, b). This nucleation of additional bands immediately adjacent to the original front or adjoining it, producing a complex front, was required for the minimization of the shear kink of the original band front. This angle of inclination of the moiré fringes, together with the angle of inclination of band fronts, proved that the material undergoes a sharp and well-defined shear at the band front. However, the shear strain alone at the front of the band cannot account for the decrease of the dimensions of specimen after the passage of the band and for the various angles of inclination of band fronts observed. The Lüders deformation may be regarded as due to a pure shear component at the band front which takes the material locally through the yield point and followed by a component due to a decelerating creep under constant stress which depends only partly on pure shear. This secondary creep effect explains the phenomenon of decrease of strain at the band front.

Examination of the original band fronts, and the moiré fringes related to them, proves that their deformation is due to pure shear. Indeed, moiré fringes did not present any other deformation at the original front except a distortion due to pure shear (see, for example, Fig. 4.12a–c). The elastic region contained between Lüders regions does not undergo any distortion. For example see Figs. 4.12a, b and 4.18a. Only in the final stages of the approach of Lüders fronts moving in opposite directions there appears some unstable elastic area split by rapidly moving secondary Lüders fronts joining the two oppositely moving primary fronts (Fig. 4.18a). The uniformity of the elastic field contained between approaching oppositely moving Lüders fronts can be seen in Fig. 4.18b, which shows a crossed moiré pattern. Both the longitudinal and transverse moiré patterns were taken instantaneously by superposing a crossed reference grating on the crossed specimen grating. Moiré fringes cross almost normal to each other at the elastic region, while in the plastic Lüders enclaves the patterns are heavily distorted.

From the conventional and the true stress–strain curves of the two types of materials tested the values of contraction ratio were calculated by using the following well-known relations:

$$v_c = \frac{de_t/\sigma_c}{de_l/\sigma_c} \quad \text{or} \quad v_r = \frac{d\varepsilon_t/\sigma_r}{d\varepsilon_l/\sigma_r}, \tag{4.1}$$

where e's designate conventional strains, ε's true strains, the subscript t means transverse strains, and the subscript l longitudinal strains. Moreover, σ_c and σ_r and v_c and v_r denote conventional and true stresses and contraction ratios respectively. The contraction ratio was found to remain constant over the elastic range. Its values are shown in Figs. 4.19 and 4.20 presenting the variation of contraction ratio versus true strains or stresses for the two types of materials tested.

After the upper yield point instability an increase of the transverse strain rate was measured which was greater than the longi-

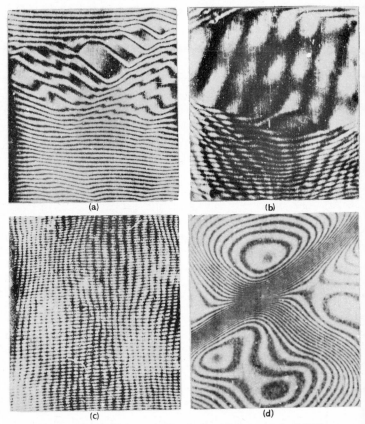

(a)   (b)

(c)   (d)

FIG. 4.18. (a) Oppositely moving Lüders fronts at a final Lüders stage where the elastic region is crossed by many unstable fronts. (b) and (c) Evolution of oppositely moving Lüders fronts and the plastic after Lüders' irregularities in the displacement field presented by crossed moiré patterns. (d) Moiré pattern showing a double V-necking of a tensile specimen.

tudinal strain rate. This phenomenon resulted in a rapid increase of the contraction ratio, the values of which quickly exceeded the value of 0·5 for the incompressible material and reached a maximum of the order of 0·65, lying at two-thirds of the Lüders strains.

After this maximum was reached the longitudinal strain rate began to increase more rapidly than the transverse strain rate, resulting in a decrease of the values of contraction ratio. The bell shape of the contraction ratio curve at the Lüders region, with a maximum exceeding the limiting value of 0·50, is due to the intrinsic instability of the material and the non-homogeneity of the strain field during the Lüders deformation. The lateral contraction ratio

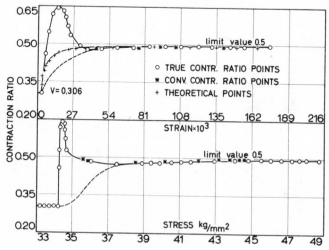

Fig. 4.19. Curves of lateral contraction ratio versus strain and stress for DIN St42.

curves at the Lüders instability region for both types of steel presented the same shape. The only difference was in the maximum value attained, which diminished with quantity of carbon contained in the steel alloys. Outside this region the contraction ratio curve continues to diminish and tends progressively to its stable branch, which is a part of a hyperbola corresponding to an ideal material with the same characteristics as the material tested, but without Lüders' instability. The stable branch of the lateral contraction ratio curve is reached at a strain much greater than the Lüders strain. The specimen continues to deform non-uniformly

up to this strain but without further nucleation of Lüders bands
Fig. 4.18d shows the moiré pattern when the specimen formed a
neck.

In conclusion it should be mentioned that this thorough study
of the instability phenomena at the yield point area of specimens
made of low carbon steel and their influence on the shape of
stress–strain and lateral contraction ratio curves was rendered

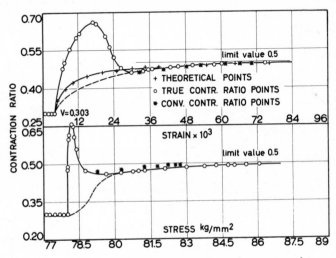

FIG. 4.20. Curves of lateral contraction ratio versus strain
and stress for USS T1.

possible only by the application of moiré methods to strain
analysis, as these enable the easy and the accurate determination
of strain values of different magnitude over relatively short lengths
at any point in a complex strain field.

A similar study was undertaken for the measurement of relative
displacements along the principal directions of flat sheet specimens
made of the same types of steel but at high temperatures where,
besides the tracing of the true stress–strain curves of the material,
the influence of temperature on the instability phenomenon due to
the nucleation and propagation of Lüders bands was studied.

The difference in the experimental procedure between the study at ambient temperatures undertaken in this section and the study at high temperatures were: (a) the crossed specimen grating was chemically etched on the surface of the specimen so as to leave a permanent impression on the surface as the photoresist burns off at elevated temperatures, and (b) the replacement of the contact differential moiré method by the image differential moiré method, because at high temperatures it was not advisable to place in contact with the hot specimen the reference grating. According to the image moiré method the image of the specimen grating was formed at unit magnification on the ground-glass screen of a still camera. The reference grating interfered with the image of the specimen grating forming a moiré pattern. This moiré pattern was photographed by a second still camera. In this manner the temperature equilibrium of the heated specimen was not disturbed by an alien body as it may by the reference grating.

For further details of the experimental results obtained by the moiré method the reader is referred to the paper by Theocaris and Koroneos.[6]

[6] Theocaris, P. S., and Koroneos, E., *Proc. Am. Soc. Test. Mater.* **64,** 747 (1964).

CHAPTER 5

Moiré Patterns
formed by Remote Gratings

5.1 Transmitted image moiré patterns of slope contours of the sum of principal stresses

In the previous chapters various moiré techniques have been described where either two similar gratings were in contact or the image of the one grating was superposed with the second grating to form a moiré pattern which gave the overall picture of the displacement field of the deformed grating. In this chapter a different type of moiré methods will be discussed where the rulings of the one grating are projected on to the second grating. The rays of projection of one grating before interfering with the second grating are distorted by the topographic irregularities of a transparent specimen under plane stress. The stressed two-dimensional specimen has its lateral surfaces distorted due to Poisson's ratio effect and is therefore transformed from a thin plate with parallel lateral faces to a lens with a variable focal length.

The moiré technique described in this section was developed by Theocaris and Koutsambessis[1] and utilizes the experimental arrangement shown in Fig. 5.1. A parallel beam of monochromatic green light passes through a reference grating RG with a principal direction parallel to the $O\xi$-axis. RG is approximately 30 mm from the front surface of the specimen Sp and has a frequency of 500 lines per inch. A camera lens L_2 is placed at a distance l_2 from RG. The image of RG is formed at unit magnification on the

[1] Theocaris, P. S., and Koutsambessis, A., *J. Sci. Instrum.* **42**, 607 (1965). See also Theocaris, P. S., *Exp. Mech.* **5** (11) 365 (1965).

ground-glass screen of the camera by adjusting l_2 to be $2f_2$. Moiré fringes are formed by placing a grating MG of the same pitch as RG in contact with the ground-glass screen. MG is in its correct position when the moiré pattern is a uniformly bright field. This position is called the zero-fringe position. The ground-glass screen is removed during the experiments and a planoconvex lens is placed beyond MG to concentrate the light beam and increase the brightness of the pattern. The moiré pattern is photographed by a second camera. The method is called the *transmitted image moiré method*.

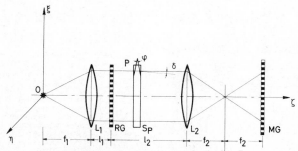

Fig. 5.1. Schematic diagram of the optical system.

If the two gratings and the faces of the unloaded specimen lie in parallel planes, the initial fringe pattern formed depicts the irregularities of the surfaces of the specimen. With specimens prepared from a Perspex sheet, which normally presents a satisfactory flatness, the initial moiré pattern shows only fractions of a fringe.

If MG is displaced from the zero-fringe position, a number of equidistant fringes parallel to the lines of the gratings will appear, which must be subtracted algebraically from the loaded pattern. The introduction of such a pattern facilitates the determination of the slope distribution in cases where the partial slope of a specimen along a particular direction varies slowly.

When the specimen is loaded, a light ray parallel to the ζ-axis, impinging at a generic point P of the specimen, is deviated through an angle δ in the $O\xi\zeta$ plane, since the faces of the specimen are no longer parallel. Moiré fringes appear on MG. Their spacing is directly related to the rate of change of slope of the surfaces of

(a)   (b)

FIG. 5.2. (a) Moiré patterns of partial slope contours along the principal axes of a perforated strip subjected to uniaxial tension for the case of a zero initial fringe arrangement (method A). (b) Moiré patterns of partial slope contours for the same specimen and loading mode when the specimen grating is approached to or receded from the reference grating in relation to its zero initial fringe position (method B).

the specimen. Figure 5.2 shows the slope contour patterns of a perforated strip subjected to uni-axial tension. Figure 5.2(a) shows contour patterns of constant partial slope for directions parallel and normal, respectively, to the direction of the applied load, MG is at the image of RG (method A). The patterns shown in

Fig. 5.2(b) are for the same loading of the specimen but with MG displaced from the image of RG approaching to or receding from the lens L_2 (method B). While the partial slope contours can be derived from all patterns, with the latter patterns detailed mapping is much easier owing to the smaller distance between adjacent fringes.

Approximate analysis based on thin-lens theory shows that the fringe order is proportional to the angle δ, the factor of proportionality depending upon the distance from RG to the centre of the specimen. Moreover, if φ is the acute angle in the $O\xi\zeta$ plane between the deformed faces of the specimen at P, it is valid that

$$\varphi = \frac{\delta}{\mu - 1}, \tag{5.1}$$

where μ is the index of refraction of the material of the specimen. Also

$$\varphi = \frac{\partial h}{\partial \xi}. \tag{5.2}$$

From the theory of elasticity it follows that

$$\varepsilon_z = \frac{\Delta h}{h} = -\frac{v}{E}(\sigma_1 + \sigma_2), \tag{5.3}$$

where v is Poisson's ratio, E the modulus of elasticity, $(\sigma_1 + \sigma_2)$ the sum of the principal stresses, and Δh is the change in thickness at P due to the applied load.

Differentiating eqn. (5.3) with respect to ξ yields that

$$\frac{\partial \Delta h}{\partial \xi} = -\frac{vh}{E} \frac{\partial(\sigma_1 + \sigma_2)}{\partial \xi}. \tag{5.3a}$$

From eqns. (5.2) and (5.3a) it is deduced that

$$\varphi = -\frac{vh}{E} \frac{\partial(\sigma_1 + \sigma_2)}{\partial \xi}. \tag{5.3b}$$

where φ now represents the acute angle corresponding to two successive moiré fringes. Then, if the fringe order is expressed by N, relation (5.3b) may be written as

$$N = k \frac{\partial(\sigma_1 + \sigma_2)}{\partial \xi}, \tag{5.4}$$

where ξ is distance measured in the plane of the RG and perpendicular to its rulings and k is a constant ($k = -vh/E$). Hence the moiré fringe pattern due to a pair of gratings with their rulings parallel to the η-axis yields the slopes of the sum of the principal stresses parallel to the ξ-axis.

Rotation of the line grating about the ζ-axis through an angle of 90° yields the partial slope contours of the sum of principal stresses along the η-axis. In problems of plasticity these partial slope contours give the variation of the lateral strain ε_z. The partial slopes along any other mutually perpendicular directions, n and t, may be found from the relations

$$s_n = \frac{\partial h}{\partial n} = \frac{\partial h}{\partial \eta} \cos \vartheta + \frac{\partial h}{\partial \xi} \sin \vartheta, \tag{5.5a}$$

$$s_t = \frac{\partial h}{\partial t} = \frac{\partial h}{\partial \xi} \cos \vartheta - \frac{\partial h}{\partial \eta} \sin \vartheta, \tag{5.5b}$$

where ϑ is the angle between the n- and η-directions.

Since the fringe order depends on the distance between the specimen and the reference grating complete evaluation of the slope distribution necessitates the accurate measurement of these quantities for an *absolute calibration*. Since the method is intended to supplement the photoelasticity method, where the difference of principal stresses is evaluated all over the field, a *relative calibration* is much easier.

It is known that along free boundaries one of the principal stresses is zero. Hence, along any such boundary the slope distribution can be found by graphical differentiation of the distribution of the sum of principal stresses. If the boundary is parallel to a principal axis it directly yields the calibration constant of either family of lines. Along an oblique or curved boundary, relations

(5.5) may be used. If the applied load is the same for both orientations of the grating the proportionality factor remains the same, thus making calibration easier.

The two photographs of moiré patterns shown in Fig. 5.2b were taken at different loads. In order to correlate the information

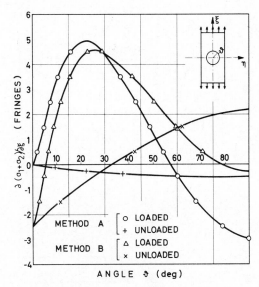

Fig. 5.3. Partial slope distribution parallel to ξ-axis at the rim of the hole of a perforated strip subjected to uniaxial tension.

derived from these photographs and to evaluate the calibration constant, the moiré pattern depicting the partial slope contours along the η-axis was multiplied by the ratio of the two loads.

Figures 5.3 and 5.4 exhibit the partial slope distribution with respect to the ξ- and η-axes along the boundary of the hole. In both figures the partial slope distribution along the same boundary was plotted for the case of an initial moiré pattern with the specimen unloaded (method B). When the unloaded pattern is subtracted from the corresponding loaded pattern, for the case of method B the difference coincides with the distribution of the method A, showing that both techniques are equally dependable. Method B has the advantage over method A of yielding more accurate information in regions of small slope variations.

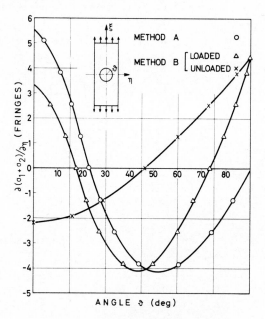

FIG. 5.4. Partial slope distribution parallel to η-axis at the rim of the hole of a perforated ship subjected to uniaxial tension.

The calibration constant was evaluated at the rim of the hole by applying eqn. (5.5b). The slope distribution for the same case was determined analytically by Howland.[2] Figure 5.5 shows the slope distribution along the boundary of the hole evaluated from Figs. 5.3 and 5.4 and also evaluated analytically by using the results of Howland. In the same figure the distribution of the sum of princi-

[2] Howland, R. C. J., *Phil. Trans. Roy. Soc.* **A 229**, 49 (1930).

pal stresses was plotted. The coincidence of the experimental with the analytical results is remarkable.

Figure 5.6 shows the slope distribution along the η- and ζ-axes as determined experimentally and analytically. Again comparison of the two groups of data shows the validity of the experimental method.

In order to evaluate the sensitivity of the method, the local shift

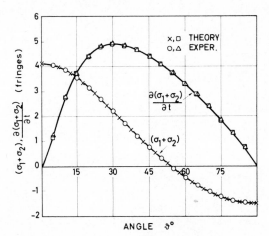

FIG. 5.5. Partial slope distribution along the tangent to the rim of the hole of a perforated strip subjected to uniaxial tension.

Δu of the image of the reference grating due to refraction at the specimen through an angle δ must be determined. This is given by

$$\Delta u = \delta . l,$$

where l is the optical distance from the reference grating to the centre of the specimen. Substituting for ϑ from eqn. (5.1) it is valid that

$$\Delta u = l(\mu - 1) \Delta\varphi.$$

In the present experiment $l = 33$ mm and $\mu = 1.5$ and the frequency of the reference grating was equal to 500 lines per inch

(20 lines per mm). If one fringe can be located to within one-fifteenth of the interfringe spacing, then

$$\Delta u = \frac{1}{300} \text{ mm}$$

and
$$\Delta \varphi = 2 \times 10^{-4} \text{ radians.} \tag{5.6}$$

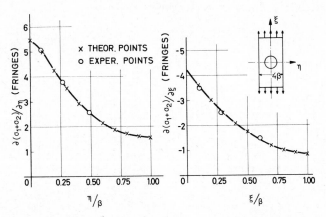

FIG. 5.6. Partial slope distributions along η- and ξ-axes of of a perforated strip subjected to uniaxial tension.

Since the sensitivity of the method is governed by the fineness of the gratings, it may be applied to the measurement of small elastic deformations as well as to the measurement of large plastic deformations.

The accuracy of the method was estimated to be of the order of 3 per cent. Among other factors, the accuracy depends on the thickness of the specimen in relation to the pitch of the reference grating. Beyond certain thicknesses the accuracy begins to fall off, but these thicknesses lie outside the range of specimens loaded in a state of plane stress.

5.2. Multisource moiré patterns of slope contours of the sum of principal stresses

In the previously described transmitted image moiré method the light rays, which pass through the reference grating, are distorted by the specimen before forming moiré fringes with the MG. In the so-called multisource moiré method the deformed specimen acts as a lens and distorts the already-formed moiré pattern by two gratings placed at some distance from each other. The light used in this case is strongly incoherent so that the reference grating, which is placed in close proximity with the light diffuser, forms a multiple light source, hence the name of the technique which was introduced by Theocaris and Koutsambessis.[3]

Consider an idealized arrangement where a monochromatic diffuser MS illuminates an amplitude line grating RG which is placed in contact with the ground-glass of the diffuser and consti-tutes the multisource grating (Fig. 5.7). A second grating, SG, is placed parallel to RG at distance a from RG having a pitch equal to $p(1 - \lambda)$, where p is the pitch of RG. The quantity λ is taken positive in order to form a real moiré pattern in the foreground of the diffuser.

In tracing the course of the light beams passing through RG and SG in Fig. 5.7, it is convenient to use a reference system which is symmetrical to both gratings. As such the plane parallel to the gratings and passing through their mid-distance was chosen. If the two gratings have their rulings parallel they form a moiré pattern at a distance l from $O\xi$. The moiré magnifications related to the reference grating M_r and to the specimen grating M_s are given by relations

$$M_r = \frac{f}{p} \quad \text{and} \quad M_s = \frac{f}{p(1 - \lambda)}. \tag{5.7}$$

From triangles ABC and AEF it can be deduced that

$$\frac{f}{p(1 - \lambda)} = \frac{l + a/2}{a}. \tag{5.8}$$

[3] Theocaris, P. S., and Koutsambessis, A., *Exp. Mech.* **8** (2) 87 (1968).

Similarly, from triangles CAD and CEF it can be shown that

$$\frac{1}{\lambda} = \frac{l + a/2}{a}. \tag{5.9}$$

Therefore

$$f = \frac{p(1-\lambda)}{\lambda}. \tag{5.10}$$

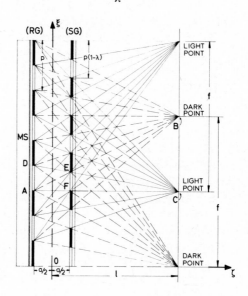

FIG. 5.7. Formation of multisource moiré patterns for $RG = p$ and $SG = p(1-\lambda)$.

This is the well-known relation yielding the interfringe spacing f in terms of the pitches of the gratings [eqn. (2.12)] and is independent of distance a.

The moiré magnifications M_r and M_s are given as

$$M_r = \frac{1-\lambda}{\lambda} \quad \text{and} \quad M_s = \frac{1}{\lambda}. \tag{5.11}$$

In the case of greatest interest, where λ is infinitesimal, the magnifications may be taken equal, i.e.

$$M = M_r \cong M_s.$$

The location of the moiré fringes from the mid-surface between the gratings is given by the distance l, which can be readily found from relations (5.8) or (5.9) together with eqn. (5.10), and is

$$l = \frac{a}{2} \cdot \frac{2-\lambda}{\lambda}. \tag{5.12}$$

Introducing the moiré magnifications M_r and M_s in relation (5.12) it is found that

$$l = a\frac{M_r + M_s}{2}, \tag{5.13}$$

and for $M_r = M_s$

$$M = \frac{l}{a}. \tag{5.14}$$

Equation (5.14) gives the moiré magnification as the ratio of the distance l of the specimen from the mid-distance of gratings and the distance a of the gratings.

If the amplitude gratings have a 50 per cent transmittance, the moiré patterns present equal opaque and light areas and there is no sharp separation between fringes. These are similar to the moiré patterns formed by two gratings superposed face to face, but they are located at a distance l from the mid-distance plane of the two gratings. It is clear from Fig. 5.7 that when λ is subtracted from p and therefore $p_r > p_s$, the image of moiré pattern is formed at the foreground of SG and is real. In the case where $p_r < p_s$ the image of the moiré pattern is formed behind the multisource RG and is virtual.

In the limiting case where fringes are formed by two identical gratings ($\lambda = 0$) both the distance l and the interfringe spacing f become infinite. Therefore, the behaviour of the pair of gratings RG and SG is analogous to a positive lens.[4]

[4] McCurry, R. E., *Appl. Phys.* **37** (2) 407 (1966).

The distance l behaves like an image distance of a positive thin lens, while p is the object distance and $p(1-\lambda)$ the focal length. The focal distance varies between 0 and $+\infty$, since p and $p(1-\lambda)$ are both taken as positive. Thus when a positive value of λ tends to zero and p tends to $p(1-\lambda)$, the distance l tends to plus infinity. If λ is negative and tends to zero from negative values, the distance l tends to minus infinity. For p tending to infinity the distance l tends to $a/2$. Finally, for p tending to zero, l tends to $-a/2$. The analogy is not complete since in the case of moiré fringes the

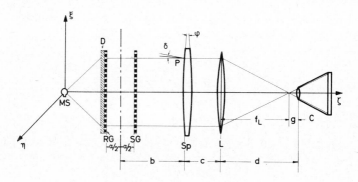

Fig. 5.8. Schematic diagram of the optical system for the multisource moiré method.

pattern is restricted to the region between $-a/2$ and $+a/2$, whereas in the case of a thin positive lens the image is restricted to lie outside the space between focal points.

The above developed theory assumes that both gratings are viewed from infinity. If the gratings are identical and are viewed through a convex lens L placed at a distance c from the specimen (Fig. 5.8), the system of lens plus gratings is equivalent to two gratings viewed from infinity with the specimen grating having an effective pitch smaller than the true pitch p_s. The moiré pattern is then formed at the focal plane of lens L and is real. The purpose of inserting the convex lens L in the optical path is solely to reduce the length of the optical bench.

In order to determine the location of moiré fringes formed by the two gratings *RG* and *SG* for different distances *a* and to prove the validity of relations (5.10) and (5.12), the following experimental arrangement was used. Two line gratings were mounted in parallel planes on a sliding table, which provided a variable distance *a* between the gratings, which was measured by a micrometer to within 0·01 mm, while excluding any relative angular displacement of the gratings.

A camera was placed at a distance of about 10 ft from the planes of the gratings. The camera with its diaphragm fully open was focused on the plane on which the fringes were formed. By measuring the distance between an object on the bench, which was in focus with the fringes, and *SG*, the fringe location was determined. The actual fringe density in the plane of formation of fringes was derived from the fringe density on the glass and the magnification factor of the camera.

For these experiments glass-plate gratings of about 500 lines per inch were used. It was found that for this line frequency the fringes were formed only at certain distinct increments of distance *a*. This phenomenon is attributed to optical interference and diffraction effects.

The results obtained from experiments with gratings of slightly different pitch proved that:

(a) The interfringe spacing is independent of the distance *a* (air gap) and it is solely a function of the difference in pitch given by relation (5.10).

(b) The fringe location for various distinct positions of the specimen grating satisfied fairly well eqn. (5.12). The discrepancy between the value of the distance *l* of the fringe location derived from relation (5.12) and the average experimental value measured by a series of different positions was of the order of 5 per cent.

In the limiting case of two identical gratings it was found that the fringes were formed at distinct increments of distance *a* and were located an infinite distance away from the camera. Furthermore, it was found that the fringe density increased by an integer for every additional increment of distance a.

In addition, different values of fringe density on the ground glass of the camera were obtained from lenses of different focal length. The fringe density decreased with increasing focal length and was independent of the distance between the specimen grating and the camera.

The above observations clearly indicate that in the limiting case of identical gratings the fringe density is solely a function of the viewing solid angle.

The experimental arrangement for the determination of slope contours is shown in Fig. 5.8. A monochromatic light source with a ground glass constituted the diffuser. An RG of 200 lines per inch was placed in contact with the ground glass. The SG, having the same frequency, with its lines parallel to the RG, was placed at a distance $a = 20$ mm. The lines of both gratings were oriented either parallel to the ξ-axis or parallel to the η-axis. The transparent specimen Sp was placed at a distance $b = 50$ mm from the mid-plane of the gratings. A lens L was set at a distance $c = 300$ mm from the specimen. A camera C was placed at a distance d from the lens L and recorded the moiré pattern formed by the inter-ference of the two gratings. Since the moiré pattern is formed in a plane different to that of the specimen and at a distance l from $O\xi\eta$, in order to have both specimen and fringe pattern focused on the camera back, the camera must be located at such a distance and the diaphragm stop chosen such that the depth of focus of the lens covers the distance between specimen and the fringe pattern planes.

If the lateral faces of the specimen were plane and parallel to the surfaces of the gratings, no distortion appeared in the moiré pattern formed by the interference of the two gratings. The density of the moiré fringes was constant and independent of the relative position of the specimen, the gratings and the lens L (independent of distances b and c). It depended on the relative position of the camera (distance d) and the distance a between gratings. There was a position of the camera where all moiré fringes of the background of the field disappeared. For $p_r = p_s$ the zero fringe position coincided with $g = 0$, i.e. when the camera lens was placed at the focus of lens L. Approaching to or receding the camera from this position formed additive or subtractive moiré patterns, the density of which increased with increasing distance g.

Any irregularity of the surface of the specimen worked like a lens and distorted the image of moiré patterns formed by the reference and specimen gratings. Indeed, when a ray of light parallel to ζ-axis impinged at a generic point P of the one face of the specimen, which formed an angle φ with its corresponding point at the other face of the specimen, it was deviated through an angle δ in the plane, since the faces of the specimen were no longer parallel to the gratings. Approximate analysis shows that the fringe order N of the moiré pattern formed by the deformed specimen is directly proportional to the angle δ. The factor of proportionality depends on the distance a and the focal length f_L of lens L. In addition, if φ is the acute angle in the $O\xi\zeta$ plane between the deformed faces of the specimen at P, relations (5.1)–(5.4) are valid. Then the moiré pattern formed by a line grating placed in a parallel plane and at some distance from a multiple light source yields the partial slope contours of the sum of the principal stresses of a loaded specimen placed at the optical axis of the system.

Figure 5.9 shows the moiré patterns of partial slopes along the principal axes of a perforated strip subjected to uniaxial tension. The moiré patterns shown in this figure are identical with the patterns given in Fig. 5.2 for the same specimen when viewed under the optical arrangement of the transmitted image moiré method. The coincidence of the patterns and the results proves the equivalence of the two methods.

5.3. The slit source and grating moiré method in plane stress problems

An interesting variation of the techniques described in §§ 5.1 and 5.2 is the method which utilizes, instead of two gratings, a slit source and a line grating. The experimental set-up used by

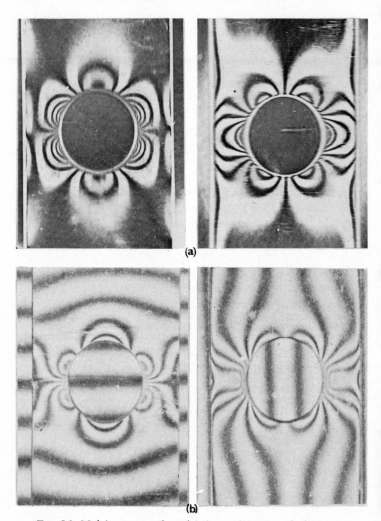

FIG. 5.9. Moiré patterns of partial slopes along the principal axes for a perforated strip subjected to uniaxial tension as obtained by the multisource moiré method. (a) Zero initial fringe arrangement. (b) With the *RG* receded from its zero initial fringe position.

Theocaris and Koutsambessis[5] during their investigations is described below.

A similar arrangement introduced by Cook[6] places the reference grating at the vicinity of the light source and the slit at the vicinity of the focal distance of the lens system of the apparatus and close to the camera. The arrangement introduced by Cook follows the set-up used by Ronchi[7] to test lenses.

Consider the idealized arrangement where the combination of a monochromatic source of light LS and a slit SL is placed at a

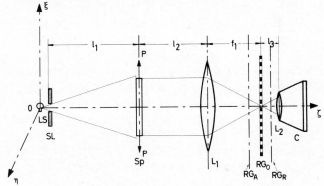

FIG. 5.10. Schematic diagram of the optical system for the slit source and grating moiré method.

distance l_1 from a thin positive lens L_1 of a focal length f_1 (Fig. 5.10), distance l_1 being greater than f_1. The system being similar to that of a projector, the rays converge to a point at a distance $l_3 = f_1 l_1 / (l_1 - f_1)$.

When a coarse amplitude grating, i.e. of a line frequency of 25 lines per inch, is inserted at the vicinity of the point of convergence of the rays of light (i.e. positions RG_A and RG_R) the light emerging from the grating has a periodic amplitude variation from zero to maximum amplitude and a frequency depending upon the focal

5 Theocaris, P. S., and Koutsambessis, A., *Strain* **4** (1) 10 (1968).

6 Cook, R. D., *Exp. Mech.* **6**, 363 (1966).

7 Ronchi, V., La prova dei sistemi ottici, *Attual. Scient.*, No. 37 (N. Zanichelli, Bologna, 1925), Ch. 9.

length of the lens L_1, the grating line frequency and the relative displacement of the grating with respect to the point of convergence. In the limiting case where the grating is placed at the point of convergence of the rays of light, i.e. at a distance l_3 from the lens, and the width of the image of the slit is equal to the width of the grating rulings, there is no obstruction of the rays light.

Let a transparent specimen be inserted between the light source and the lens at a distance l_2 from the lens, where $l_2 < f_1$. With the specimen distorted due to an externally applied load, its lateral surfaces form sections of positive or negative thin lens of a large and variable focal length. This variable lens combined with the lens L_1 causes the rays of light to converge to points different to that of l_3 depending upon the type of lens into which the specimen is transformed and its focal length. The positions of the amplitude bands are displaced following the simple magnification formulae of lens combinations.[8] By placing a camera at a distance l_4 from the grating and focusing it on the specimen, changes occurring in the patterns can be observed and recorded with a fringe background depending upon the relative position of the grating with respect to the point of convergence. As distance l_4 does not influence the sensitivity of the method it can be chosen such as to attain a reasonable magnification of the specimen on the camera back.

On the same theoretical considerations of the two previous sections it can be shown that the moiré fringes formed by this experimental set-up represent contour curves of partial slope of the sum of principal stresses along the principal direction of the rulings of the grating.

By angularly displacing the slit and the line grating about the ζ-axis through an angle of 90° the partial slope contours of the sum of the principal stresses along the second principal axis of the specimen (η-axis) are obtained. It is possible to use a pinhole and an orthogonally crossed grating and obtain simultaneously the two families of contours in one pattern. At points of high

[8] Jenkins, F. A., and White, H. E., *Fundamentals of Optics*, 3rd edn., McGraw-Hill, 1957, pp. 52–55.

gradients of the sum of principal stresses this technique may introduce difficulties in distinguishing the patterns of the contours corresponding to each grating.

In order to show the validity of the technique it was applied to the same problem of a perforated strip subjected to longitudinal tension treated in the two previous sections. The moiré patterns of the partial slope contours in two orthogonal directions corresponding to the zero-fringe position of the reference grating were absolutely identical to the corresponding patterns shown in Figs. 5.2 and 5.9 and are therefore omitted. Figure 5.11 presents

FIG. 5.11. Moiré patterns of partial slope contours along the principal axes of a perforated strip subjected to uniaxial tension.

the partial slope contours for the perforated strip when the grating is approached to the lens L_1. It can be stated from these figures that the clarity of fringes is remarkable. Moreover, the pattern obtained with a crossed grating is shown in the same figure. The quality of this pattern is satisfactory and therefore the pattern can be used for a simultaneous evaluation of slope distributions along the two principal directions.

The evaluation of the calibration constant as well as of the sum of principal stresses along any traverse of the specimen follows the same procedure described in the previous sections and therefore it does not need to be repeated.

The main differences between this set-up and that of Cook[9] is

[9] Cook, R. D., *Exp. Mech.* **5** (11) 363 (1966).

that in this set-up a diffused light illuminates the specimen while in Cook's arrangement collimated light is used. Moreover the image of the grating G in Cook's arrangement (fig. 1 of ref. 9) is projected on the slit S, while in the set-up of this method the image of the slit is projected on the grating RG. For the zero fringe position RG_o both systems are equivalent since the grating is working in this case as a simple slit. For positions of RG different from that of RG_o, as are positions RG_A and RG_R in Fig. 5.10, there is a difference between the two systems, since in this set-up the grating RG is working as a multiple slit on to which the image of slit SL is projected, while in Cook's arrangement the image of the multiple slit (grating G) is projected on slit S. This difference accounts for the difference in quality of the moiré fringes obtained by the two systems. Sharp and well-contrasted moiré fringes, obtained by this set-up, significantly contribute to the accuracy of the method.

5.4. Curvature distributions obtained by frequency modulated amplitude gratings

Consider an amplitude grating RG combined with a free of aberrations thin positive lens L_1 interposed in the path of a monochromatic light beam created by the light source LS and the diaphragm M (Fig. 5.12).

The following normalized coordinates are used,

$$\left. \begin{array}{l} \zeta = (k \sin \alpha)\, Z, \\[6pt] \zeta' = (k \sin \alpha')\, Z', \\[6pt] x = \dfrac{X}{l_0}, \end{array} \right\} \tag{5.15}$$

where Z, Z', X and l_o are the geometric distances along the ζ-, ζ'- and x-axes respectively, and l_0 is the radius of the aperture of lens L_2. Quantity k is equal to $2\pi/\lambda$, where λ is the wavelength of light used in the experiment. Angles α and α' are defined in Fig. 5.12.

If the amplitude transmission of the combination of RG and the thin lens L_1 is varying along the ζ-direction according to

$$T(\zeta) = \tfrac{1}{2}(1 + a_c \cos \omega_c \zeta),$$

then the amplitude A of the coherent wave impinging on the combination will be changed to

$$F(\zeta) = \tfrac{1}{2}A(1 + a_c \cos \omega_c \zeta) \qquad (5.16)$$

after emergence from the combination. It was assumed in this

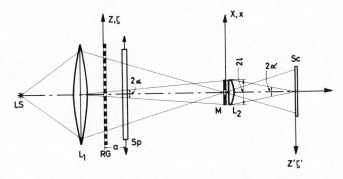

FIG. 5.12. Schematic representation of the optical element consisting of a projector and a grating.

relation for simplicity that the phase of the impinging wavefront at the moment of entering the combination is equal to zero. In these relations a_c is the amplitude of transmission and ω_c the radian frequency of light used. If the spatial variable ζ in eqn. (5.16) is replaced by the time variable t the grating RG with lens L_1 may be regarded as a carrier-wave oscillator converting the continuous component of light into a sinusoidal wave $F(\zeta)$. If the amplitude a_c of transmission of RG and L_1 as well as the amplitude A of the plane wave impinging on RG are considered for simplicity equal to unity then eqn. (5.16) becomes

$$F(\zeta) = \tfrac{1}{2}(1 + \cos \omega_c \zeta). \qquad (5.17)$$

If a specimen *Sp* placed in the optical path of the light beam passing the grating-lens combination at a distance *a* from *RG* suffers some deformation of its lateral surfaces due to an applied stress, point ζ on the grating viewed through the deformed transparent specimen will be displaced to point $\zeta + f(\zeta)$. Function $f(\zeta)$ is the signal in the problem and for reasons of simplicity is assumed to be expressed as

$$f(\zeta) = a_m \sin \omega_m \zeta, \tag{5.18}$$

where the frequency ω_m of the signal is smaller than the frequency ω_c of the carrier.

Line grating *RG* is placed in front of lens L_1 and each spectral component from *RG* is focused by lens L_1 on its focal plane *P* located at the entrance pupil of the projection lens L_2. On plane *P* a diaphragm *M* is placed, which changes the phase of some of the spectral components on *P*. The modified spectra form through lens L_2 the image of the signal on screen *Sc*. A transparent thin plate, stressed in plane stress is made of a photoelastically inert material (like Perspex), so that any other influence of the stressed material on the transmission of the wavefront except the influence due to thickness variation is excluded.

Substituting the quantity $(\zeta + a_m \sin \omega_m \zeta)$ in place of ζ into relation (5.17), it is deduced that

$$F_m(\zeta) = \tfrac{1}{2}[1 + \cos \omega_c(\zeta + a_m \sin \omega_m \zeta)]. \tag{5.19}$$

The instantaneous frequency of $F_m(\zeta)$ is given by the derivative, with respect to ζ, of the expression in brackets in eqn. (5.19) and which is

$$\frac{d}{d\zeta}[\omega_c(\zeta + \alpha_m \sin \omega_m \zeta)] = \omega_c(1 + \alpha_m \omega_m \cos \omega_m \zeta). \tag{5.20}$$

This expression shows that the frequency of the carrier wave has undergone a frequency modulation by the signal.

It may be assumed that $b = \alpha_m \omega_c < 1$, since the specimen is assumed to suffer infinitesimal deformations and therefore small shifts in the deformed image of the grating. This assumption implies that the maximum shift of a generic point in the deformed image of the grating is smaller than the wavelength of the carrier wave. Since $\omega_c > \omega_m$ it is valid *a fortiori* that

$$a_m \omega_m < 1. \tag{5.21}$$

The spectral distribution $\varphi(x)$ at the focus of the condensing lens L_2 created by a monochromatic light point source S is given by

$$\varphi(x) = \int_{-l}^{+l} F_m(\zeta) \exp(-ix\zeta)d\zeta, \tag{5.22}$$

where l is the radius of lens L_2 outside of which $F_m(\zeta)$ is assumed negligible.

The spectra $\varphi(x)$ can be obtained for the frequency modulated amplitude carrier wave by substituting eqn. (5.19) into eqn. (5.22).

If a diaphragm M is placed at the focal plane P of lens L_2 it will modify $\varphi(x)$ according to its transmittance $m(x)$. Then the spectra $\varphi'(x)$ of modified carrier after passing through the diaphragm M is given by

$$\varphi'(x) = \varphi(x)m(x). \tag{5.23}$$

The spectrum $\varphi'(x)$ will form an image through the lens L_2 on the screen Sc whose amplitude will be given by

$$F'_m(\zeta') = \int_D \varphi'(x) \exp(-i\zeta'x)dx, \tag{5.24}$$

where the domain of integration D represents the frequency range of spectrum allowed to pass through the lens L_2. Since the x-axis corresponds to the spatial frequency axis, the aperture of L_2 may be regarded as a frequency passband of the bandpass filter L_2 through which the frequency components $\varphi'(x)$ can proceed to the screen Sc. For both the source S and the lens L_2 having their centres on the optical axis all spectra within the frequency range from $-l$ to $+l$ will pass through the lens.

By cutting off some spectral components by the diaphragm M the signal can be extracted from the carrier wave.

The intensity distribution on the screen will be given by

$$I'(\zeta') = F_m'(\zeta')^* . F_m'(\zeta'), \qquad (5.25)$$

where the asterisk denotes the complex conjugate of $F_m'(\zeta')$. The amplitude $F_m'(\zeta')$ of the image of $\varphi'(x)$ through lens L_2 on screen Sc is given by eqn. (5.24) and therefore

$$I'(\zeta') = \iint_D \varphi'(x_1)\varphi'(x_2)^* \exp\left[-i(x_1 - x_2)\zeta'\right] dx_1 dx_2. \qquad (5.26)$$

If $\varphi'(x)$ is represented as the sum of line spectra, relation (5.26) can be written as

$$I'(\zeta') = \sum_{x_1} \sum_{x_2} \varphi'(x_1)\varphi'(x_2)^* . \exp\left[-i(x_1 - x_2)\zeta'\right]. \qquad (5.27)$$

In general, $\varphi'(x)$ is a complex function, but the case of particular interest is that when the imaginary part of the function is equal to zero. Therefore, if $\varphi'(x) \equiv \varphi_r'(x)$ [$\varphi_r'(x)$ is the real part of function $\varphi'(x)$], eqn. (5.27) may be written as

$$I'(\zeta') = \sum_{x_1} \sum_{x_2} \varphi_r'(x_1)\varphi_r'(x_2) \cos(x_1 - x_2). \qquad (5.28)$$

Then, if the spectra within $-\omega_c \leqslant x \leqslant \omega_c$ are cut off it may be deduced from eqn. (5.28) and the approximate values of Bessel functions needed, where only terms up to the third order were considered from their power series expansions, that

$$I'(\zeta') = \frac{b^2}{2}\left(1 + \frac{b}{2}\cos \omega_m\zeta'\right). \qquad (5.29)$$

If the diaphragm cuts off the spectra within the range $-\omega_c \leqslant x \leqslant +\omega_c$ the intensity becomes

$$I'(\zeta') \approx 2(1 + b \cos \omega_m\zeta'). \qquad (5.30)$$

In either case of eqn. (5.29) or (5.30) the image on screen Sc is related to the derivative of the signal.

If the loaded transparent specimen is located somewhere between the lenses L_1 and L_2, its deformed lateral faces constitute a variable, positive and negative thin lens, then the signal passing through the specimen expresses the partial slope distribution along an axis coinciding with the principal direction of the grating RG due to the lateral deformation of the specimen. Therefore, the image on screen Sc, which is related to the derivative of the signal, expresses the slope of the partial slope curve of the deformed body.

For a signal $f(\zeta) = a_m \sin \omega_m \zeta$, the image received on Sc yields

$$\frac{d}{d\zeta} f(\zeta) = a_m \omega_m \cos \omega_m \zeta. \tag{5.31}$$

The amplitude of the signal on Sc is given by b, as is readily derived from relations (5.29) and (5.30). The amplitude b on Sc is modulated with respect to the amplitude $a_m \omega_m$ in (5.31) by the factor ω_c/ω_m. It is always valid that $\omega_c/\omega_m \gg 1$. This is a convenient fact because it increases the sensitivity of the method and allows the measurement of deformations in the specimen with very small ω_m.

The specimen is placed at the vicinity of lens L_1. In this manner the sensitivity of the method increases considerably. It is also possible to replace the coarse grating by two fine gratings placed at some distance apart and lying in parallel planes. It was previously shown that the effect of the two gratings corresponds to a sole grating with a much coarser pitch.

Figure 5.13 shows the moiré pattern of partial slope contours along the transverse axis of a perforated strip subjected to a uniaxial tensile load along its longitudinal axis. This pattern is formed by the combined action of two identical gratings RG_1 and RG_2 placed at the vicinity of lens L_1 in parallel planes and at some distance apart. A second pattern is superposed to the first one, which yields the curves of equal slope of the partial slope contours and it is characterized by the large and sparse moiré fringes. The first pattern which yields the partial slope contours is formed by dense moiré fringes. It is worth while mentioning that the fringes of the second pattern appear where the slopes of the first pattern expressing the partial slope contours are abruptly changing direction. The fringe order N of the pattern yielding the slope contours of the partial slope distribution in the specimen expresses therefore the partial curvatures of the sum of principal stresses existing at the deformed specimen according to the relation

$$N = C\frac{\partial^2(\sigma_1 + \sigma_2)}{\partial x^2}. \tag{5.32}$$

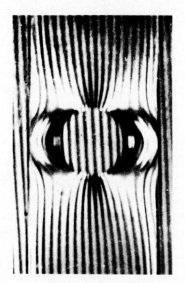

FIG. 5.13. Moiré pattern of partial slope contours along the transverse axis of a perforated strip subjected to a uniaxial tensile load along its longitudinal axis. Superposed to this pattern is the pattern of equal slopes of the partial slope contours given by the first pattern.

A similar relation holds for the pattern obtained with the grating angularly displaced by 90° where the second partial derivative is taken with respect to the y-axis. Both components contribute to the evaluation of the total second derivative of the sum of principal stresses along an arbitrary traverse of the specimen.

By a double integration of the total second derivative of $(\sigma_1 + \sigma_2)$ it is possible to evaluate the distribution of the sum of principal stresses at the interior of the field of a two-dimensional specimen planely stressed. Since the method is intended to complement photo-elastic information yielding the difference of principal stresses all over the plane stress field, constant C can be calibrated at a free boundary of the specimen where both the difference and the sum of principal stresses are coincident. The values of the sum of principal stresses derived from this method combined with the values of difference of principal stresses yielded by the photo-elastic method allow the separation of principal stresses all over the field.

In order to check the validity of the technique to yield a partial curvature contour map of the deformed specimen a mechanical differentiation of the partial slope map of the specimen was executed. The mechanical differentiation method was introduced to moiré applications by Dantu[10] and utilized by Duncan and Sabin,[11] Ligtenberg,[12] Post,[13] Heise,[14] and Parks and Durelli.[15] This technique consists in superposing two patterns of identical slope contours with a constant displacement between the patterns and along their principal direction. Figure 5.14 shows a moiré pattern where the fringes represent lines of constant difference between slopes caused by the displacement Δx. The incremental differences of slope along the direction of displacement approach the corresponding derivatives when the magnitudes of displacement are approaching zero. This reduction of the necessary displacement is only feasible in cases where the initial moiré patterns are dense. Otherwise, large displacements are required in order to yield a few second-order fringes.

In order to determine the sensitivity of the method, an amplitude grating is assumed in position RG, whose transmittance is

[10] Dantu, P., Lab. Centr. Ponts Chaus., Publ. No. 57–6 (1957.)

[11] Duncan, J. P., and Sabin, P. G., E. Mech. 5 (1) 22 (1965).

[12] Ligtenberg, F. K., Proc. Soc. Exp. Stress Anal. 12 (2) 93 (1954).

[13] Post, D., Exp. Mech. 5 (11) 368 (1965).

[14] Heise, V., Exp. Mech. 7 (1) 47 (1967).

[15] Parks, V. J., and Durelli, A. J., Trans. ASME J. Appl. Mech. 33E (4) 901 (1966).

expressed by eqn. (5.16). It is also assumed that the variation of position of each ruling from its initial position due to the distortion of the image of the grating through the deformed specimen is given by the simple wave relation

$$f(\zeta) = a_m \sin \omega_m \zeta. \qquad (5.33)$$

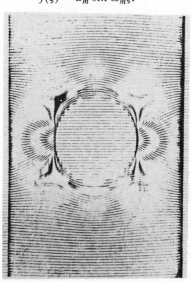

FIG. 5.14. Moiré pattern of constant differences between partial slopes along the transverse axis derived by a constant displacement Δx of the pattern yielding the corresponding partial slope contours.

The intensity of the image of the deformed grating on screen Sc is given by the relation in compliance with the previous theory

$$I'(\zeta') = 2(1 + a_m \omega_c \cos \omega_c \zeta'),$$

if the spectra in the region $-\omega_c < x < +\omega_c$ are assumed to be cut off by the diaphragm.

If it is further assumed that a 10 per cent variation of the intensity of the image $I'(\zeta')$ can be estimated then

$$\alpha_m \omega_c = 0 \cdot 10.$$

But it may be also written that

$$\alpha_m \omega_c = 2\pi A_m F_c, \tag{5.34}$$

where A_m is the amplitude variation of the position of two successive rulings of the distorted image of RG and F_c the line frequency of RG. For a grating with $F_c = 20$ lines per mm it is deduced from eqn. (5.34) that

$$A_m = \frac{0 \cdot 10}{2\pi F_c} = 0 \cdot 8 \, \mu. \tag{5.35}$$

The quantity $A_m \cong 1 \cdot 0 \, \mu$ expresses the smallest detectable variation in the image of two successive rulings in RG due to the distortion of its image when viewed through the deformed specimen. If the distance between the grating RG and the specimen is $a = 2$ in., the angle $\Delta\varphi$ under which a chord of 1μ is viewed from the specimen is

$$\Delta\varphi = 0 \cdot 35 \times 10^{-6} \text{ rad.} \tag{5.36}$$

Since the pair of gratings utilized in Fig. 5.13 are equivalent to one much coarser grating, the sensitivity of the technique in this figure is much lower and this explains the sparsity in curvature contours.

5.5. Moiré patterns formed by inclined gratings

While in the previous sections the applications of moiré patterns formed by remote line gratings placed in parallel planes have been discussed, in this section a geometric obstruction theory of the interference of two mutually inclined line gratings is given. The angle subtended by the planes of the gratings is considered as the main parameter and its influence on the spatial location of the

moiré patterns is studied. This method was developed by
Theocaris.[16]. Two interesting cases will be considered when the
rulings of the gratings are either normal (case A) or parallel
(case B) to the edge of the dihedral angle formed by the planes of
the gratings.

Let the plane Oxy of an $Oxyz$ Cartesian system of coordinates

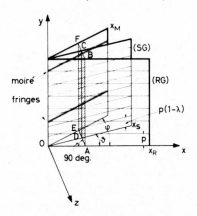

FIG. 5.15. Formation of
moiré patterns by inclined
gratings with their rulings
normal to the edge of their
dihedral angle.

coincide with the reference grating and let the plane of the specimen grating form an angle ϑ with the reference plane (Fig. 5.15).
Both gratings are amplitude-type parallel line gratings and the
reference grating is placed in contact with the ground-glass screen
of a monochromatic diffuser. Consider first the case when the
rulings of the gratings run normal to the y-axis. The pitch of the
rulings in RG is p, while the pitch of the rulings in SG is $p(1-\lambda)$,
where λ is a small positive quantity. The assessment of λ-positive
is not restrictive, since, for λ-negative, the real moiré pattern
formed in the foreground space of SG will become virtual and will

[16] Theocaris, P. S., *Brit. J. Appl. Phys.* (*J. Phys.* D), Ser. 2, **1** (7) 678 (1968).

be formed in the background space of *RG*, in complete analogy with the theory of simple positive lenses.

Moreover, it is assumed that both gratings are viewed from infinity on an optical axis coinciding with z-axis. Consider a section plane *ABCD* normal to the reference plane Ox_Ry at a distance x from origin O and parallel to y-axis. The section of the reference and specimen grating planes by the plane *ABCD* will yield two families of equispaced slits and bars lying on parallel lines and at a distance $AD = x \tan \vartheta$ apart. The interference of the two families will form moiré dots lying on the *ABCD* plane and on the *EF* line, which is parallel to the lines *AB* and *CD*. The distance of moiré line from the mid-distance of the two dot lines is given by relation (5.12),

$$l_m = \frac{AD}{2}\left(\frac{2-\lambda}{\lambda}\right). \tag{5.37}$$

The distance AD is given by

$$AD = x \tan \vartheta.$$

Hence,
$$l_m = \left(\frac{2-\lambda}{2\lambda}\right)x \tan \vartheta. \tag{5.38}$$

Therefore the distance of the moiré line *EF* from the reference plane is

$$l = x\left[\frac{2-\lambda}{2\lambda} \tan \vartheta + \tfrac{1}{2} \tan \vartheta\right],$$

which yields
$$l = x\frac{\tan \vartheta}{\lambda}. \tag{5.39}$$

By displacing the *ABCD* plane along the x-axis it can be readily shown that the moiré pattern lies in a plane inclined with respect to the reference grating plane whose slope is

$$\left[\frac{\tan \vartheta}{\lambda}\right].$$

Then the angle φ is given by

$$\varphi = \arctan\left(\frac{\tan \vartheta}{\lambda}\right). \tag{5.40}$$

The moiré fringes lying on the moiré plane $Ox_m y$ run normal to the y-axis. The moiré fringes are equispaced parallel fringes having an interfringe spacing f given by

$$f = \frac{p(1-\lambda)}{\lambda}.$$

This relation is identical with the expression yielding the interfringe spacing of moiré fringes formed by two line gratings lying on parallel planes [eqn. (5.10)] and indicates that the moiré pattern is not affected by the inclination angle between the planes of the gratings.

Consider now the case when the rulings of the two gratings run parallel to y-axis. Let the Oxz plane be normal to the reference plane and the traces of the reference grating and the specimen grating on the Oxz plane be the lines Ox_R and Ox_S. Consider the interference of the kth order bar of RG and the $(k+1)$th order bar of SG. The straight line AB joins the boundaries of these bars (Fig. 5.16).

The indicial equation of the rulings of the reference grating is given by

$$x = kp. \tag{5.41}$$

The indicial equation of the rulings of the specimen grating can be readily found from simple geometric relations and is expressed by

$$z = \left[x \cot \vartheta - \frac{lp}{\sin \vartheta} \right], \tag{5.42}$$

where k is an indexing parameter running over some subset of the real integers and l is another indexing parameter defining the families of bundles of lines centred at each boundary of slits and bars in the reference grating and passing through the boundaries of all neighbouring slits and bars of the specimen grating. Then, while k defines the order of each bundle starting at the boundaries of slits and bars in RG, the integer l defines the order of line in each bundle (Fig. 5.17).

The coordinates of points A and B are, respectively,

$$\left.\begin{array}{ll} x_A = kp & x_B = (k+l)p(1-\lambda)\cos\vartheta \\ z_A = 0 & z_B = (k+l)p(1-\lambda)\sin\vartheta \end{array}\right\} \quad (5.43)$$

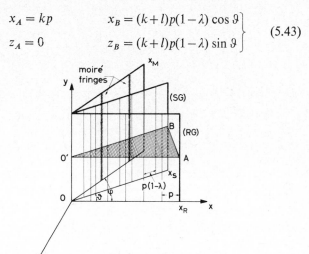

FIG. 5.16. Formation of moiré patterns by inclined gratings with their rulings parallel to the edge of their dihedral angle.

The equation of the line AB on the plane $Ox_R z_S$ is given by

$$[(k+l)p(1-\lambda)\sin\vartheta]x + [kp - (k+l)p(1-\lambda)\cos\vartheta]z$$
$$- k(k+l)p^2(1-\lambda)\sin\vartheta = 0. \quad (5.44)$$

Relation (5.44) also expresses the family of bundles formed for various values of k with l as parameters. In order to determine the equation of the envelope of lines AB joining the boundaries of the bars of the gratings it is necessary to differentiate the left-hand side of the parametric equation (5.44) with respect to the parameter k

and to equate this derivative to zero. Differentiation of eqn. (5.44) yields

$$x = \frac{(2k+l)p(1-\lambda)\sin\vartheta - [1-(1-\lambda)\cos\vartheta]z}{(1-\lambda)\sin\vartheta}. \qquad (5.45)$$

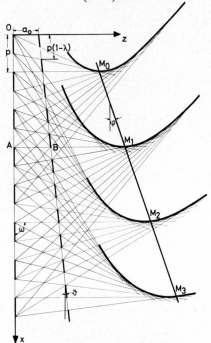

Fig. 5.17. Geometry of formation of moiré fringes in space by two inclined gratings.

Introducing the value for x into (5.44) and solving for x and z it is deduced that

$$\left.\begin{array}{l} x = \dfrac{(k+l)^2 p(1-\lambda)\cos\vartheta - k^2 p}{l}, \\[3mm] z = \dfrac{(k+l)^2 p(1-\lambda)\sin\vartheta}{l}. \end{array}\right\} \qquad (5.46)$$

Relations (5.46) are the parametric equations of the envelopes of the bundles of lines AB (Fig. 5.17). It is clear from the parametric equations of the envelopes of lines AB and from Fig. 5.16 that:

(a) When the pitch of SG is smaller than that of RG the moiré pattern is real and is located in the foreground space of SG.

(b) When the pitch of SG is larger than that of RG the moiré pattern is virtual and is located in the background space of SG.

However, it is possible to define points or lines along each envelope curve or surface where the light intensity of the moiré fringes becomes minimum. These are the points or lines for which the tangents to the envelopes are normal to the plane of RG and therefore parallel to the viewing direction. The points of the envelope curves are anticipated when sections $Ox_R x_s$ normal to the y-axis are considered. If the reference and specimen planes are contemplated the envelopes become three-dimensional surfaces and the points of minimum intensity turn into lines parallel to the y-axis. The moiré pattern formed by these points or lines lie on a curve or surface, which is called the effective moiré pattern.

In order to define the position of effective moiré pattern it is necessary to seek the points or lines at the family of envelope curves or surfaces for which the tangents to the envelopes form an angle $\omega = 90°$ with the Ox axis. For the right-angle triangle OAB it is valid that

$$k = \frac{l(1-\lambda)\cos\vartheta}{1-(1-\lambda)\cos\vartheta}. \qquad (5.47)$$

Introducing this value for k into relations (5.46), the coordinates of the point on the envelope may be found whose tangent subtends an angle of $90°$ with the Ox axis. The coordinates of this point are

$$\left.\begin{aligned} x &= \frac{pl(1-\lambda)\cos\vartheta}{[1-(1-\lambda)\cos\vartheta]}, \\ z &= \frac{pl(1-\lambda)\sin\vartheta}{[1-(1-\lambda)\cos\vartheta]^2}. \end{aligned}\right\} \qquad (5.48)$$

Taking successive integral values of l gives the coordinates of adjacent points of minimum light intensity M_0, M_1, M_2, etc., of the effective moiré pattern.

It can be readily found from the equation of straight lines that the points of the envelopes whose tangents are normal to the reference plane lie on a straight line which passes through the intersection of the traces Ox_R and Ox_s of the reference and specimen gratings. In the case where, instead of the envelope curves which lie on the Oxz plane, the envelope surfaces are considered, the points of the surfaces whose tangent planes are normal to the reference plane lie on a plane passing through the y-axis. The slope of the line or the plane of effective moiré pattern is given by the relation

$$\tan \varphi = \frac{\tan \vartheta}{[1 - (1 - \lambda) \cos \vartheta]}. \tag{5.49}$$

Angle φ expressed by relation (5.49) defines the $Ox_M y$ plane along which the effective moiré pattern is formed. The moiré fringes on this plane appear as equidistant parallel fringes when viewed from infinity.

If the moiré fringes are viewed from a direction normal to the reference grating and under a certain viewing angle, they appear with a variable interfringe spacing since the viewing plane sees the moiré envelopes at different points than the points corresponding to the effective moiré pattern. This is the reason why the intensity of moiré fringes is variable with the fringes formed at points far from their effective points being blurred and with a reduced illumination.

There is a certain angle φ_0 between the moiré and the reference planes for which no moiré pattern is formed. This is the case where the projections of the bars of the reference grating are covering the slits of the specimen grating. It can be deduced from simple trigonometry of the projected bars that

$$(1 - \lambda) \cos (\varphi_0 - \vartheta) = \cos \varphi_0. \tag{5.50}$$

Relation (5.50) rearranged yields

$$\tan \varphi_0 = \frac{[1-(1-\lambda)\cos \vartheta]}{(1-\lambda)\sin \vartheta}. \tag{5.51}$$

Angle φ_0, defined by relation (5.51), yields the limit of inclination of the moiré fringe plane for which the moiré pattern vanishes.

In order to check the theory developed and prove the validity of relation (5.49), a series of tests with inclined gratings was executed. A pair of line gratings of a line frequency of 8 lines per millimetre was mounted in planes forming a variable dihedral angle. The reference grating was in contact with a diffuser consisting of a bright monochromatic light source and a ground-glass screen and remained fixed during the experiments. The specimen grating with $\lambda = 0.005$ was mounted on a pivoting frame. This frame could rotate about an axis either parallel or normal to the direction of the rulings of the grating. Angular displacements ϑ of SG were of the order of 3° and could be measured by a protractor. A camera was positioned at a distance of 2 m from the planes of the gratings and could be oriented so that the principal plane of its lens could form any angle with the plane of RG. The camera with its diaphragm fully open was focused on the plane on which the moiré fringes were formed.

The fringe location was also determined by a ground-glass screen which could pivot about the edge of the dihedral angle of the two gratings. Since the specimen gratings tested were always of a higher frequency than the reference grating, the moiré patterns were real and were formed on the screen in the foreground of the specimen grating. The screen was angularly displaced about the edge till the moiré pattern appeared in focus all over its plane. The angle subtended between the ground-glass screen and the reference grating yielded angle φ of eqn. (5.49). In the case where the frequencies of the specimen gratings were lower than the frequency of the reference grating, the moiré patterns were virtual and were formed in the background of it. In this case the ground-glass screen was removed and the orientation of the camera lens yielded the angle of rotation of the moiré plane.

By changing the distance between the camera and the gratings the solid angle under which the camera was viewing the gratings was changed. In this manner the influence of the direction of viewing on the moiré pattern with respect to the plane of gratings was studied. Figure 5.18 shows the moiré patterns formed by two inclined gratings the rulings of which are either normal (Fig. 5.18a) or parallel (Fig. 5.18b) to the edge of the dihedral angle of the two gratings. While the moiré fringes in the pattern of Fig. 5.18a are equispaced, the fringes in the pattern of Fig. 5.18b are of a variable interfringe spacing. This phenomenon is explained by taking into consideration that the camera was placed close to the moiré plane and focused on the moiré envelopes at points different from those corresponding to the effective moiré pattern. Figure 5.18c shows the moiré pattern in the case where the two gratings are viewed from the angle φ_o of zero fringe pattern. The fringes appearing at the upper and lower sides of the figure are again due to the close proximity of the camera to the gratings.

In order to satisfy the condition of placing the camera at infinity and to reduce the distance between gratings and camera, a convex thin lens was interposed between the gratings and the camera. Figures 5.18d–f show the moiré patterns formed by two mutually inclined gratings with their rulings either parallel (Fig. 5.18d) or normal (Fig. 5.18e) to the edge of the dihedral angle of the grating planes in the case where a convex lens is interposed between the gratings and the camera. In this case the moiré patterns are formed by parallel and equidistant fringes for both positions of the gratings. Figure 5.18e shows, besides the primary moiré pattern formed by mechanical interference of the two gratings, a beat phenomenon derived from optical diffraction and interference of the primary line gratings. A similar phenomenon appears in Fig. 5.18f where the rulings of the inclined gratings are parallel to the edge of the dihedral angle of the grating planes. This beat phenomenon also explains the staggering shape of moiré fringes at the vicinity of the extrema of the wave.

In order to check the validity of eqn. (5.49) a series of tests with a pair of inclined gratings of a line frequency of 20 lines per milli-

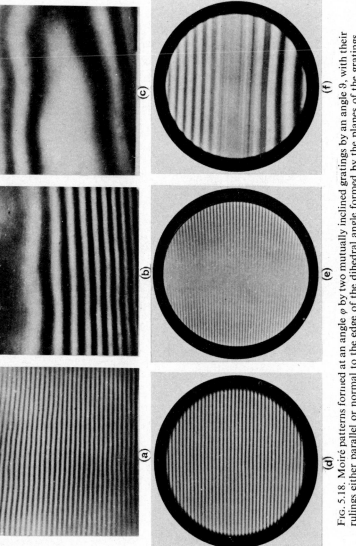

Fig. 5.18. Moiré patterns formed at an angle φ by two mutually inclined gratings by an angle ϑ, with their rulings either parallel or normal to the edge of the dihedral angle formed by the planes of the gratings.

metre was executed with $\lambda = 0.012$ in./in. The angles φ were measured between the ground screen and the reference grating for various values of angle ϑ subtended between the planes of the two gratings. Figure 5.19 shows the theoretical and experimental

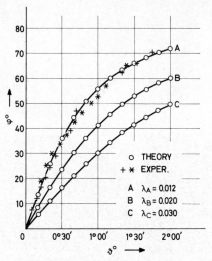

Fig. 5.19. Comparison of theoretical and experimental values for angle φ of inclination of the moiré plane for various angles ϑ of inclination of the specimen grating.

results for angles ϑ up to $2°$. Since the measurement of angle φ was made by simple means (protractors) the coincidence of the experimental results with the theoretical values may be considered as satisfactory. In the same figure the theoretical curves $\varphi = f(\vartheta)$ were also traced for $\lambda = 0.020$ and $\lambda = 0.030$ in./in.

CHAPTER 6

Surface Topology
by Moiré Patterns

IN THIS chapter the moiré methods employed for experimentally detecting and defining deflections, slopes, curvatures, twist of plane surfaces of constructions subjected to external loading, will be considered. The suitable preparation of a physical surface for the application of each of the moiré techniques is beyond the scope of this chapter, and information concerning the technology of surface preparation can be found in the appropriate bibliography. Since in moiré applications the quality of surface preparation is not so critical as in optical interferometry, fine grinding or electro-polishing is satisfactory in most cases. A lapping process may be undertaken in cases where the flatness of a surface is of particular interest. Fine grinding for high polymer surfaces is recommended since lapping is very difficult if not impossible. In some plastics like Perspex the as cast surface of Perspex sheets is in many cases satisfactory.

The fundamental mechanism of the moiré effect may be used to record lateral displacements w and its derivatives. The lateral displacement of a deformed body may result either from Poisson's ratio effect or from a deflection of the body.

In the case of lateral displacements due to Poisson's ratio effect the displacements w are given by the isopachics that is the geometric loci of equal sum of principal stresses. In this case the displacements w are of comparable order of magnitude with the in plane u- and v-displacements, the measurement of which has been described in Chapters 2–4. The methods described in Chapter 5 are concerned with the measurement of the slopes of isopachics in an elastic field.

In this chapter a moiré method of direct measurement of the lateral displacement w due to Poisson's ratio effect will be introduced.

For flexed bodies the lateral displacements w become large compared to in plane displacements u, v. In this case the construction acquires large deflections, slopes and curvatures. Hence the second part of this chapter will be devoted to methods dealing with the measurement of these quantities in flexed bodies.

6.1. Isopachic patterns by the shadow moiré method

An important application of moiré fringes in metrology is to provide topographic information about irregularities of a surface. Thin film interferometry has been adopted as a usual means of mapping much smaller topographic irregularities of a surface than those measured by moiré methods. The first example of interference fringes of equal thickness is the so-called Newton's rings which are of historical interest in connection with Newton's views on the nature of light.[1] These are formed by the air gap between a convex spherical surface of a lens and a plane glass plate in contact. The fringes are concentric circles about the point of contact. With normal incidence the condition for a dark fringe is

$$r = \sqrt{(mR\lambda)}, \qquad (6.1)$$

where r is the radius of the fringe, R is the radius of curvature of the lens surface, m is an integer indicating the fringe order, and λ the wavelength of the monochromatic light beam. Relation (6.1) indicates that the radii of the dark fringes are proportional to square roots of the positive integers and, therefore, Newton's rings constitute excellent zone gratings. In fact this was the way of preparation of zone gratings shown in Fig. 3.13. Cornu[2] also used thin-film interferometry to map an anticlastic surface of a bar in pure bending and measure Poisson's ratio of the material.

[1] Newton, I., *Opticks*, 2nd Book, Part 1 (based on the 4th edn., 1730), Dover Publ. Inc., 1952.

[2] Cornu, A., *Compt. Rend. Acad Sci. Paris* **69**, 333 (1869).

Interference isopachic patterns of transparent or metallic specimens loaded in plane stress were obtained by several experimenters. Favre,[3] in 1927, used the Mach–Zehnder interferometer and a transparent specimen with optically flat surfaces. Interference is obtained by reflection on the front or the rear surface of the specimen. Similar applications were made by Fabry,[4] who silvered both surfaces of the specimen. Tesar[5] and, separately, Frocht[6] used a reference flat transparent surface and obtained interference between this surface and the initially flat metallic specimen surface. Interesting isopachic patterns were obtained by Sinclair[7] using transmitted light, a Mach–Zehnder or Jamin interferometer and two identically loaded Bakelite specimens. Another set-up yielding isopachics by transmitted light was introduced by Bubb.[8] Bubb's method used only one transparent specimen and a Michelson type interferometer. All the above-mentioned methods have not found a broad application because of the difficulty of obtaining optically flat surfaces on the specimen and, on the other hand, the necessity of expensive, elaborate and very sensitive instrumentation and of considerable experimental skill in order to obtain the interference patterns.

The use of the moiré effect considerably simplifies the optical technique of obtaining isopachic patterns. In this case the specimens need not have optically flat surfaces. Apparently, the first investigators to use the moiré phenomenon for the measurement of thickness variations in plane specimens were Dose and Landwehr.[9] They proposed a procedure for originally non-flat specimens in which the optical interference fringes obtained before and after loading, when interfered, created a moiré pattern of the isopachics. Mesmer[10] applied the same method to a transparent

[3] Favre, H., *Schweitz. Bauztg.* **90**, 291 and 307 (1929).

[4] Fabry, C., *Compt. Rend. Acad. Sci. Paris* **190**, 457 (1930).

[5] Tesar, V., *Revue Opt.* **11**, 97 (1932).

[6] Frocht, M. M., *Proc. Fifth Int. Cong. Appl. Mech., Cambridge, Mass., U.S.A.*, 1938, p. 221.

[7] Sinclair, D., *J. Opt. Soc. Am.* **30**, 511 (1940).

[8] Bubb, F. W., *J. Opt. Soc. Am.* **30** (7) 297 (1940).

[9] Dose, A., and Landwehr, R., *Naturwiss.* **36**, 342 (1949).

[10] Mesmer, G., *Proc. Soc. Exp. Stress Anal.* **13** (2) 21 (1956).

thin specimen with nearly plane surfaces. When an incident beam of monochromatic light was partly reflected from the front surface and partly from the rear surface of the transparent specimen under external load, the two reflected parts of the light beam interfered and created an interferogram, which was used as a loaded grating. A similar interferogram was taken with the specimen unloaded. Photographic superposition of these patterns gives a moiré pattern corresponding to the isopachic distribution of the loaded specimen. Drouven[11] and, separately, Post[12] applied Mesmer's technique in special problems. Post[13-15] introduced a new large-field interferometer for the determination of isopachics consisting of three reflecting surfaces in series contained in one half-mirror and a double mirror. Frappier[16] modified Post's interferometer by introducing a third independent half-mirror treated only on the one surface, which replaced one of the two treated surfaces of the second mirror. Pirard[17] studied theoretically the phenomenon of formation of moiré fringes by using parametric relations of various families of curves. He applied also the method for the determination of isopachics in plane–stress fields by using transparent specimens.

Theocaris[18] used thin film interferometry for the evaluation of the lateral displacement of transparent two-dimensional specimens subjected to plane stress. The interference pattern was formed by the air gap between the front reflective surface of the specimen and a reference surface of an optical flat placed in contact with the specimen. The same method was applied by Oppel and Hill.[19] Duncan et al.[20] developed also a similar interferometric technique

[11] Drouven, G., Doct. Diss., Washington University, Saint Louis, U.S.A., 1952.

[12] Post, D., *Proc. Soc. Exp. Stress Anal.* **12** (1) 99 (1954).

[13] *Ibid.* **12** 191 (1954).

[14] *Ibid.* **13** (2) 119 (1955).

[15] Post, D., *J. Opt. Soc. Am.* **44** (3) 243 (1954).

[16] Frappier, E., *Analyse des Contraintes, Mém. GAMAC* **2** (8) 29 (1957).

[17] Pirard, A., *Analyse des Contraintes, Mém. GAMAC* **5** (2) 1 (1960).

[18] Theocaris, P. S., *J. Mech. Phys. Solids* **11** (3) 181 (1963).

[19] Oppel, G. U., and Hill, P. W., *Exp. Mech.* **4** (2) 206 (1964).

[20] Duncan, J. P., *Proc. Inst. Mech. Engrs. London*, **176**, 379 (1962).

suitable for quantitative analysis of flexural problems in opaque specimens polished by a simple lapping technique. All these methods are based on optical interferometry and they use the moiré effect for deducing the lateral displacement data from the interferograms. They are, therefore, beyond the scope of this book, and they will not be described in any further details.

In fields of fairly large lateral displacement it is possible to more readily extract all the information obtainable from an interferogram by using the moiré effect engendered by the interference of a

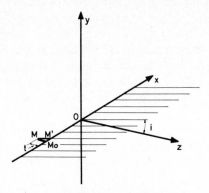

FIG. 6.1. Geometry of formation
of shadow moiré fringes.

reference grating and its shadow on the matte reflecting surface of the specimen. The method was introduced by Theocaris.[21]

The principle of the method is presented in Fig. 6.1. The Oxy plane denotes the plane of the undeformed model and coincides with or is parallel to the reference grating plane. Oz is the axis normal to this plane. Oblique incidence of a parallel beam of white light under a chosen angle i may result in a moiré pattern formed by interference of the rulings of the reference grating and their shadows or reflections on the deformed specimen.

A generic point M_0 of the specimen may be considered to be on

[21] Theocaris, P. S., *J. Sci. Instrum.* **41**, 133 (1964). See also Theocaris, P. S., *Exp. Mech.* **4** (6) 153 (1964); *ibid.* **7** (7) 289 (1967).

the Ox axis without losing generality, if the collimated light beam is taken obliquely incident to this axis. Point M_0 will be displaced to point M, due to the lateral deformation of the specimen. If $M_0M = t'$, then the change of thickness of the specimen will be $2t'$. Superposition of a reference grating containing an array of parallel rulings of a frequency of 500 lines per inch results in an initial moiré pattern representing the surface distortion of the specimen. After application of the load on the specimen a new moiré pattern appears due to the lateral deformation of the specimen. Let us examine the influence of the normal displacement t' of the specimen at point M_0 on the moiré pattern. Suppose that all normal displacements of the surface are small relative to the depth of the field, so that the image produced on the photographic plate is the principal projection of the grating on the deformed surface of the specimen. If M_0M is the normal displacement of the specimen surface, then point M' of the reference plane is projected on point M.

When interference of rulings and their shadows on the matte surface of the specimen is sought for the formation of moiré fringes, the displacement $M'M_0$ of the shadow of a generic ruling is equal to

$$M'M_0 = t \tan i. \tag{6.2}$$

If p denotes the pitch of the master grating, p' the pitch of the projection of the shadow of the rulings on a plane parallel to the reference grating, and if f indicates the interfringe spacing and n is the number of grating spaces between two adjacent moiré fringes, then from Fig. 6.2 it is valid that

$$AC = (n-1)p, \quad AD = (n-1)p' = np, \quad CD = p, \tag{6.3}$$

for small angles i.

Distance $BD = t$, which represents the height of the air gap between RG and Sp at the points where two successive fringes occur, may be determined from triangle BDC and expressed by relation

$$t = \frac{p}{\tan i}, \tag{6.4a}$$

which, for an angle of viewing ρ, becomes

$$t = \frac{p}{\tan i + \tan \rho},\qquad(6.4\text{b})$$

Thus the moiré pattern formed by the lines of a reference grating and their oblique shadows on the surface of the specimen yields its thickness variation in terms of the pitch of the reference grating.

Evaluation of the lateral deformation of a loaded specimen

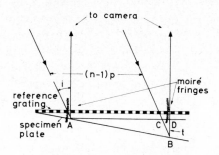

Fig. 6.2. Interference produced by the lines of a grating and their shadows formed by an obliquely incident light beam on the matte surface of a deflected plate.

allows the determination of the mean normal stress distribution in a plane elasticity field or the immediate calculation of the lateral normal strain in a plane–stress plasticity problem.

It is known that the lateral strain ε_z is connected with the sum of principal stresses in an elastic plane stress field by relation

$$\varepsilon_z = \frac{\Delta h}{h} = -\frac{v}{E}(\sigma_1 + \sigma_2),\qquad(6.5)$$

where v is Poisson's ratio and E the modulus of elasticity, h and Δh the thickness and the change of thickness of the loaded specimen respectively. If the isopachic fringe order is denoted by N, then

$$N = \alpha\Delta h = -\frac{\alpha v\,h}{E}(\sigma_1+\sigma_2),$$

$$N = k(\sigma_1+\sigma_2),$$

(6.6)

where k is a calibration constant defined by the corresponding isochromatic pattern and the free boundary conditions. Determination of the sum of principal stresses in combination with their difference yielded by the photoelastic method allows the separation of the principal stresses all over the elastic field. Evaluation of the lateral strain ε_z and its increments after each step of loading in a plane stress plasticity problem may simplify the relations expressing the components of strains and stresses and their increments and considerably facilitate the long and tedious calculation procedure. Therefore it allows the solution of concrete plasticity problems which remain unsolved because of the prohibitive amount of calculations by the pointwise photoelastic method of normal and oblique incidence.

The optical arrangement used for the experiments was simple. A white-light spot source containing an incandescent lamp and a diaphragm was used to produce a stigmatic beam of light. A monochromatic sodium lamp can also be used, although the contrast of fringes obtained with this lamp is poorer than that obtained with white light. The angle of incidence of light was regulated by a mirror so as to produce the maximum number of moiré fringes without losing much in contrast. Convenient angles of incidence i producing dense and sharp fringe patterns were found to be of the order of $10°$. Where the specimen was made of transparent material, its front surface was polished with zinc sulphide and sprayed with a thin and uniform layer of an aluminium paint to become matte. Since the front surface was the reflecting surface, there was no restriction in the selection of accidentally refractive materials. Thus the same specimen can be used, without any inconvenience, after its use in photoelastic analysis. Since Perspex satisfactorily fulfils the requirement of flatness of the specimens at least over sufficiently large areas of the customary sheets, the same specimen was used for the tracing of isoclinics and later for obtaining the isopachics. In the case of opaque metallic materials, their front surface must be polished in order to reflect the incident light and sprayed with an aluminium paint in order to form a matte surface.

The reference glass plate containing the grating was mounted with the emulsion side facing the reflecting surface of the specimen. The influence of the glass plate curvature must be considered because there is a small displacement of the specimen when photographed unloaded and loaded. The plate curvature contributed a small error in the measurements. Actually, if the reference plate is selected to present a deviation from a true plane less than the ruling pitch of the grating, the error introduced remains insignificant. Photographic plates with such precision in flatness are easy to obtain.

The reference grating was reproduced photographically from an originally ruled plate. The frequency of the rulings of the reference grating was chosen equal to 500 lines per inch, and the direction of the rulings was chosen to coincide with one of the principal axes of the specimen. This resulted in a remarkable refinement of the moiré fringes formed and in a symmetric pattern which considerably simplified the subsequent evaluation of data.

As an illustration, the experimental method was applied to a problem already solved theoretically and experimentally by other known methods in order to compare the theoretical and experimental results and thus attain an assessment of the accuracy of the method. But the applicability of the experimental method can be extended to solve any kind of plane elastic problem, as well as to determine ε_z strains normal to the surface of plane stress plasticity problems.

A strip with two semicircular grooves subjected to uniform tension was used for illustration. The diameter of the grooves was chosen equal to half the width of the strip. The specimen was made from a Perspex plate of thickness $t = 10$ mm, selected from an almost flat part, having a width $w = 160$ mm and a groove radius $r = 40$ mm. The specimen was made sufficiently long in order to secure a uniform stress distribution in the strip away from the grips.

Since the pitch of the reference grating was very large in comparison to the measured variations of thickness of the specimen, the Perspex specimen was loaded in its rubbery state, where it presents large elastic deformations. In this manner the number of moiré fringes was considerably increased. The specimen was immersed in an oil bath and loaded whilst at the rubbery-state temperature. It was kept loaded until the temperature dropped to the ambient and, thus, stresses and strains were frozen in the specimen. In this manner, the specimen sustained large elastic deformations which yielded sufficiently dense isopachic patterns. The surface of the specimen was coated with an aluminium paint to form a matte surface. After loading the specimen was placed in the parallel white-light beam. The reference glass plate was superposed on the specimen.

The image of the reference grating was formed on the surface of the specimen. A camera was placed at an oblique axis to the surface of the specimen to photograph the moiré fringes. The angle of the camera axis was chosen such that the camera could collect a maximum of the reflected light from the surface of the specimen on the ground glass. This angle was of the order of 10°. Since the reference plane rides on the high points of the specimen there may be an eccentricity in the support of this plane. Therefore, an interference pattern will be formed with the specimen unloaded. This pattern, which is called original pattern, is first photographed. A new pattern is formed with the application of the load which is called loaded pattern. The loaded pattern, which is shown in Fig. 6.3, contains, besides the original pattern and the pattern due to the lateral deformation of the specimen, another pattern due to the inevitable rotation of the reference plane. Since the reference plane is riding on the high points of the specimen and since there is a variation of their elevation due to loading, some rotation of the reference plane may be expected. The irregularities observed on the surface of the unloaded specimen were insignificant and much lower than the unit of measure related to the half-pitch of the grating. The original patterns presented only fractions of fringes all over the surface of specimens and were insensitive to any change in position of the reference plane.

Observation of both sides of the loaded specimen showed a difference of fringes along the longitudinal axis of the specimen. Hence, the specimen was reheated to its rubbery-state temperature after the loaded pattern was photographed, and left unloaded to

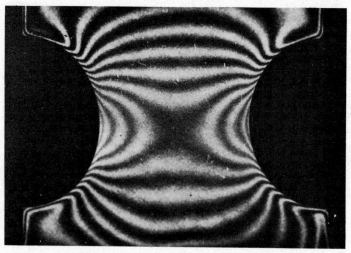

FIG. 6.3. Loaded moiré pattern of shadow fringes for a symmetrically grooved specimen subjected to uniaxial tension.

cool to the ambient temperature. This unloading procedure relieved all fringes except those due to the flexure of the specimen during loading. These fringes were taken into account for the evaluation of isopachics along the longitudinal axis of the specimen.

After the original and the loaded patterns were photographed, it was necessary to establish in what direction normal to the fringes their order increases or decreases. This procedure is simple, provided care is taken to properly recognize peaks, valleys and necks in the topographic configuration of the patterns. Valleys and peaks of a harmonic surface must lie always at the boundaries of the model and they can be readily recognized since they give patterns with concentric curves. The convexity or concavity of the surface may be determined from the corresponding photo-elastic pattern since the shape and spacing of isopachic curves at free boundaries must be proportional to the shape and spacing of the corresponding isochromatic pattern.

In order to evaluate the variation in thickness along any traverse section or boundary of the specimen, it is necessary to trace a diagram of the variation of thickness of the unloaded specimen. This diagram is traced with an arbitrary scale because the thickness variation corresponding to one interference fringe is unknown, and with an arbitrary shifting from zero level because the absolute fringe order is also unknown. The diagram of the loaded configuration of the specimen is traced in the same manner with the same fringe scale and respecting also the already established sequential direction of fringe variation. The difference of the two diagrams yields the thickness variation traced to an arbitrary scale. The scale of the moiré fringes can be evaluated either by an absolute calibration or by a relative calibration at a free boundary.

The absolute calibration can be performed by accurately measuring the pitch of the reference grating under a microscope and the angle i of incidence of light. The procedure of relative calibration is easier and may be performed on a free boundary. The sum of principal stresses at a free boundary coincides with their difference. This difference may be determined by the corresponding photo-elastic pattern previously obtained by the same specimen or a similar one under similar conditions of loading. The calibration constant k in relation (6.6) is defined as the number of moiré fringes corresponding to the distance between two consecutive isochromatic fringes and may be determined at a free boundary. Evaluation of the calibration constant k in this manner allows the conversion of the moiré pattern of isopachics to the same scale as the corresponding isochromatic pattern. In order to determine the arbitrary shifting of the moiré pattern of isopachics, it is necessary to correlate the value of an isopachic passing through an arbitrary point at a free boundary with the corresponding value of the isochromatic pattern. Either free corners or points of intersection of the principal axes of the specimens with free boundaries are usually selected as such points.

The calibration constant was evaluated at the rim of the groove with reference to the values derived from the existing experimental solution.[22] Figures 6.4 and 6.5 show the variation of the sum of

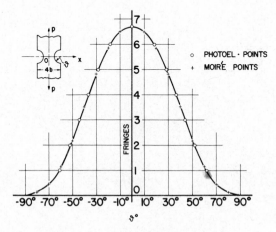

FIG. 6.4. Distribution of the sum of principal stresses at the rim of the grooves of a symmetrically grooved strip subjected to uniaxial tension.

principal stresses at the rim of the groove, the minimum section, and the longitudinal axis of the specimen as evaluated by this method. The comparison of the experimental results with a previous experimental solution based on photo-elastic coatings[22] shows a remarkable agreement and proves the accuracy of the method.

The validity of the so-called shadow moiré method appears in the results of the problem treated and the comparison with the results obtained by other experimental methods. The correlation

[22] Theocaris, P. S., *Trans. ASME J. Appl. Mech.* **29** (84) 735 (1962).

between these results is highly satisfactory and therefore substantiates the applicability of this method to practical problems. Moreover, the moiré method presents many advantages over the more conventional experimental methods presently available. These methods are either tedious point-by-point methods or else require highly elaborate equipment and intricate techniques, and

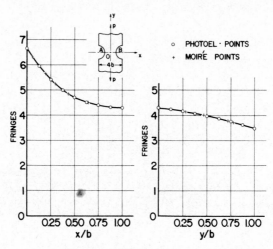

FIG. 6.5. Distribution of the sum of principal stresses at the minimum section and the longitudinal axis of a symmetrically grooved strip subjected to uniaxial tension.

some of them absolute flatness of the specimens used. The simplest of them necessitate transparent specimens and they require, for a complete photoelastic analysis, two specimens identical in shape subjected to identical stress distributions. A second specimen is needed because the material used for interferometric measurements cannot be birefringent, for a unique velocity of light is required through the whole thickness of the specimen. In general, all these methods require many precautions to be taken and considerable experimental skill in order to obtain reliable isopachic patterns.

The moiré method is a simple and reliable method. It requires neither monochromatic light nor elaborate equipment. A simple white-light parallel beam and a camera is needed. The reference grating may be easily obtained on a flat glass plate through a simple photographic-reduction procedure. It may be applied to any kind of specimen made either of photoelastic material or of metallic opaque material. The moiré fringes formed are low-order fringes and therefore the interference patterns are sharp. The isopachic fringe order at each point is the difference between the fringe orders of the loaded and the original pattern. Thus initial non-uniformity of the specimen thickness is taken into account and consequently has no effect on the accuracy of the measurements. Thus the method does not necessitate optically flat surfaces.

The method is easily adjusted and does not necessitate any kind of isolation from vibrations. It may be easily applied to dynamic problems. Its validity becomes very significant in the analysis of plane stress plasticity problems with metallic specimens, where the evaluation of transversal strain considerably facilitates the calculations.

6.2. The oblique shadow moiré method for measurement of deflections in plates

The method described in § 6.1 and applied to the measurement of lateral displacements of two-dimensional transparent specimens loaded in plane stress may be readily adopted to measure deflections of flexed plates. The shadow moiré method was introduced by Theocaris[23] for the measurement of deflections in flexed plates.

The technique used was similar to the technique utilized for the case of transparent flat specimens in plane stress. In this case the deformed front surface of the two-dimensional specimen was replaced by the upper surface of the flexed plate. The plate specimen may be made either from Perspex, when the as cast surface of the plate is satisfactory for the formation of moiré fringes or from

[23] Theocaris, P. S., *Proc. Int. Symp. Shell Struct.*, *Warsaw 1963*, North Holland Publ. Co., Amsterdam, 1965 p. 887.

any material, which, in this case, is preferable to be the same as the material of prototype, and the surface must be ground to a certain degree of flatness and then sprayed and covered by a thin layer of an aluminium oxide to become matte reflecting.

The reference grating glass plate was riding on the high points of the upper surface of the plate. Since the deflections of the plate are much larger than the lateral displacements of a two-dimensional specimen in plane stress, the line frequency of the grating may be smaller than in the first case. A grating with a frequency of 200–400 rulings per inch gives dense moiré patterns with the plate flexed in the elastic region.

For a relatively flat Perspex or a metallic plate the original moiré pattern presents only fractional orders of moiré fringes. If the total number of fringes in the loaded pattern is of the order of 10–15 fringes, neglecting the original fringe distribution due to inherent irregularities of the surface introduces an error of the order of 2–5 per cent and may therefore be neglected. For a more accurate evaluation of the deflection distribution the reference grating was mounted on an adjustable tripod which allowed orienting the reference grating at will relatively to the position of the specimen and rendered the grating plane independent from any rotation of the specimen during loading. Moreover, the original pattern with the plate unloaded was photographed and subtracted from the loaded pattern to yield the moiré pattern due to the deflection of the plate.[24]

The relative deflection w between two points of the surface of the plate is given by

$$w = \frac{p}{\tan i}, \qquad (6.7)$$

where p is the pitch of the reference grating and i is the angle of incidence of the light beam. Relation (6.7) is the same as relation (6.4a). In the case of viewing the plate under an angle ρ relation (6.4b) is valid.

For the evaluation of the deflection distribution along any

[24] Theocaris, P. S., and Hazell, C. R., *J. Mech. Phys. Solids* **13**, 281 (1965).

traverse of the specimen the diagrams of deflection of the unloaded and loaded pattern along the traverse are traced. These diagrams, although both are traced to the same scale, have an arbitrary displacement from zero level since the absolute fringe order in the specimen is unknown. The scale of moiré fringes can be readily determined from an absolute calibration by measuring the pitch of the reference grating, the angle i of incidence of the light beam and eventually the angle of viewing ρ.

For the evaluation of the moments and shears it is necessary to determine the modulus of rigidity of the plate. Since the elastic constants for Perspex vary with temperature and humidity, these constants must be determined during each test. This can be done by measuring the deflection of a calibration plate of known dimensions under a known load for which a theoretical solution exists. A clamped circular plate of radius a, loaded with a uniform pressure, may be used as a calibration plate. The deflection of the centre of the plate is given by

$$w_c = \frac{qa^4}{64D_c}. \tag{6.8}$$

This relation yields the rigidity D_c of the calibration plate when the maximum deflection w_c is measured for a certain applied pressure q. The rigidity D of the specimen made of the same material as the calibration plate can be readily deduced from

$$D = D_c(t/t_c)^3, \tag{6.9}$$

where t and t_c denote the thicknesses of the specimen plate and the calibration plate respectively. For the evaluation of moments the second derivatives of deflection with respect to x and y must be determined from the deflection curves and the value of Poisson's ratio for the material.

As an illustration the method was applied to two different problems. The first problem is the case of a circular plate of diameter d, simply supported at the outside circumference and loaded with a uniformly distributed load over a concentric circle of diameter a. The ratio a/d was taken equal to 0·206. Since the geometry and the loading of the plate were symmetric the deflection variation was independent of angle ϑ. Figure 6.6 shows the original and loaded moiré patterns for three loading steps. All

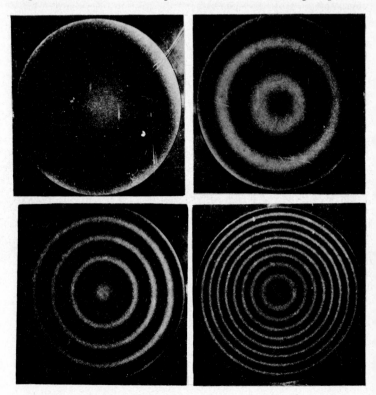

FIG. 6.6. Original and loaded moiré patterns for three loading steps of a circular plate simply supported at its rim and flexed by a load applied at its centre.

moiré patterns present a radial symmetry. Figure 6.7 presents the deflection distribution along a radius of the plate and Fig. 6.8 gives the comparison of the experimental results with theory. The remarkable coincidence of both groups of results shows the accuracy of the experimental method.

While in the first illustrative example a flexed plate in the elastic region was considered, in the second example cases of plastically deformed plates are studied. Rhomboid plates in anticlastic bending were used for the study of the shape of the initial and subsequent yield loci of the material of the plate. As material an

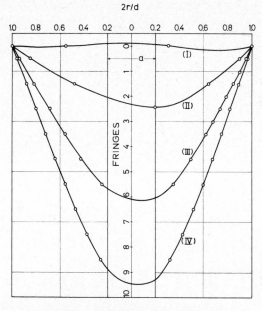

FIG. 6.7. Deflection distribution along a radius of a circular plate simply supported and flexed by a concentrated central load.

aluminium alloy was used under commercial designation 6061-T-651, which presented substantial work hardening. Each corner of the plate was prepared with a small tab to provide means of applying anticlastic bending through a specially designed jig placed on a conventional testing machine. The anticlastic loading of the rhomboid plates was maintained by placing concentrated downward loads at two opposite corners and upward at the other two corners. These loads replaced the twisting moments which produced pure bending and were uniformly distributed along the sides of the rhombus. A slight indentation at the tabs located hard, steel balls through which the external loads were applied.

The polished surface of each plate was sprayed with an aluminium–chrome enamel to form a matte reflecting surface. After adjusting the apparatus to form a symmetrical loaded moiré pattern the plates were loaded up to the yield stress of the material

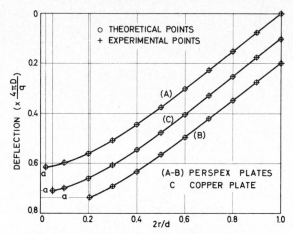

Fig. 6.8. Comparison of experimental and theoretical results for the deflection distribution along a radius of two Perspex and one copper circular plate simply supported and centrally loaded.

and the moiré pattern of the deflection contours was photographed after each loading step. Figure 6.9 shows the moiré patterns of deflection contours in anticlastic bending of two types of rhomboid plates.

If in the rhomboid plate the coordinate axes coincide with the diagonals of the rhombus, the entire plate is in a state of pure bending, the principal moments M_1 and M_2 at the centre being

$$M_1 = M_x = -\frac{P}{2\gamma} \quad \text{and} \quad M_2 = M_y = \frac{P.\gamma}{2}, \qquad (6.10)$$

where P is the applied load at the corners and γ is the ratio of the diagonals of the rhombus. By altering the ratio γ, combinations of

FIG. 6.9. Moiré patterns of deflection contours in two types of rhomboid plates flexed in anticlastic bending.

principal bending moments can be obtained which cover the entire anticlastic quadrants of the yield locus of the material.

It was shown[25] that, when Mises' yield criterion is assumed, the principal bending moments can be expressed as functions of the

[25] Lerner, S., and Prager, W., *Trans. ASME, J. Appl. Mech.* **27**, 353 (1960).

yield moment M_0 and the ratio γ. These are given by

$$M_1 = -\frac{M_0}{(\gamma^4 + \gamma^2 + 1)^{\frac{1}{2}}} \quad \text{and} \quad M_2 = \frac{M_0 \gamma^2}{(\gamma^4 + \gamma^2 + 1)^{\frac{1}{2}}}. \quad (6.11)$$

The yield load P_0 can be determined by equating the moment of a corner load with respect to a diagonal to the total bending moment

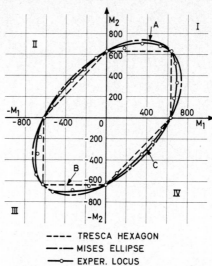

---- TRESCA HEXAGON
—·— MISES ELLIPSE
—○— EXPER. LOCUS

FIG. 6.10. Initial yield surface of aluminium 6061-T-651 obtained experimentally by using flexed plates and compared with Mises' and Tresca's yield loci.

transmitted along the diagonal. This may be readily found to be

$$P_0 = \frac{2\gamma M_0}{(\gamma^4 + \gamma^2 + 1)^{\frac{1}{2}}}. \quad (6.12)$$

When Tresca's yield criterion is assumed the following relations hold:

$$M_1 = -\frac{M_0}{(1 + \gamma^2)} \quad \text{and} \quad M_2 = \frac{M_0 \gamma^2}{(1 + \gamma^2)}, \quad (6.13)$$

and
$$P_0 = \frac{2\gamma\,M_0}{(1+\gamma^2)}. \tag{6.14}$$

Rhomboid plates of various values of γ loaded anticlastically were used to investigate the yield locus in the second and fourth quadrants. The yield load P_0 was determined from the loading machine. From the value of P_0 and relations (6.10)–(6.14) the experimental values of the yield moments for the plate, as well as the corresponding values for the cases of validity of Mises' or Tresca's yield criteria were determined. Figure 6.10 shows the initial yield surface of aluminium 6061-T-651 as obtained by experiments with plates and compared to Mises' and Tresca's yield loci. The values of the experimental points of the yield locus at the first quadrants were obtained by loading plates in synclastic bending, which was not described above. For further details of the application of the method to the study of the yield locus of various materials see ref. 24, p. 234.

6.3. Ligtenberg's photoreflective moiré method for recording slopes

Several early methods[26] for recording the changes of slope in plates have used square, orthogonal or circular and radial grids which by reflection on the polished surface of the plate gave the changes of slope. These techniques, although accurate, are cumbersome, and hence they have been outmoded by the easier moiré methods.

The moiré effect, as utilized in the study of flexed plates, may be considered as a secondary outcome of the development of slopes in the deformed plate. The partial slopes of the plate create a displacement of the reflected image of the grating on the reflective surface of the plate which interferes with the image of the initial undistorted grating and forms the moiré patterns, yielding a map of partial slope contours of the flexed plate.

[26] Anthes, H., *Dinglers Polytech. J.* 342, 356, 388, 441, 455, 471 (1906). See also Dantu, P., *Annls. Ponts Chauss.* **110** (1) 5 (1940) and **122** (3) 271 (1952). Moore. A. D., *Trans. ASME, J. Appl. Mech.* **17** (3) 291 (1950).

The method was developed by Ligtenberg[27] and was applied to various problems of flexed plates[28] as well as to two-dimensional problems of elasticity by application of the slab analogy.[29]

The apparatus used for this method consists of the specimen plate placed in a loading jig, whose front surface has been specularly reflective. At a distance d (Fig. 6.11) from the specimen a screen is placed made up of dark rulings traced on a white surface with a pitch $p = 0.0894$ in. (2.27 mm). The dark rulings are covering the half space of p. A camera is placed at the centre of the screen and views the specimen through a perforation O. With the plate unloaded the image of a certain point Q of the screen is formed at the point S on the ground-glass screen of the camera via reflection at point P of the plate. With the plate deformed, at the same point S of the screen of the camera, another point R of the screen is viewed through reflection at point P' of the specimen. The distance QR enables the determination of the slope ϑ. By making a double exposure of the undistorted image of the screen with the plate unloaded and the distorted image of the screen with the plate flexed, a moiré pattern appears on the negative showing the partial slope contours of the plate in the principal direction of the rulings of the screen. This is because the difference in the two superposed images is primarily caused by the angular slope

[27] Ligtenberg, F. K., *Proc. Soc. Exp. Stress Anal.* **12** (2) 83 (1954).

[28] Vreedenburgh, C. G. J., and Van Wijngaarden, H., *Proc. Soc. Exp. Stress Anal.* **12** (2) 99 (1954). See also: Van der Sande, G. A. F., *Ingenieur* **68** (13) 17 (1956). Loof, H. W., Stevinlaboratorium Rep. TU 13 (1956). Palmer, P. J., *Aircr. Engng.* **29** (346) 377 (1957). Bouwkamp, J. G., Delft Inst. Techn. Rep. P. IV (1952). Bradley, W. A., *Proc. ASCE. J. Engng. Mech Div.* **85** (EM4) 77 (1959). Ligtenberg, F. K., and Bouwkamp, J. G., Delft Inst. Techn. Rep. H–II (1952). Duyster, T. H., TNO weer Bouwmaterialen and Bouwconstructies, Delft, Rep. 27545 (1957). Casper, W. L., M.S. Thesis Res. Rep. Div. Struct. Engng. and Struct. Mech., Univ. of Calif., Berkeley (1960).

[29] Ligtenberg, F. K., and Loof, H. W., Stevinlaboratorium Rep. DV 5 (1957). See also: Loof, H. W. and van der Sande, G. A. F., *Selected Papers Stress Anal.* 1961, p. 20. Bouwkamp, J. G., *Proc. First Int. Cong. Mech.*, New York, 1963, p. 195. Bouwkamp, J. G., *Exp. Mech.* **4**, 121 (1964). White, E. I., M.S. Thesis Res. Rep. Div. Struct. Engng. and Struct. Mech., Univ. of Calif., Berkeley (1959). Loof, H. W., Delft Inst. Techn. Rep. TS–I (1957). Zienkiewicz, O. C., and Cruz, C., *Int. J. Mech. Sci.* **4**, 285 (1962).

changes of the plate surface, and therefore the moiré fringes are a direct measure of this change in terms of the known dimensions of the apparatus. The influence of the deflection of the plate at P is small and may be neglected.

The distance QR is known and measured by the moiré fringes. The angular deflection ϑ can be found from simple trigonometry and is

$$QR = 2d \cdot \vartheta \, (1 + b^2/d^2), \qquad (6.15)$$

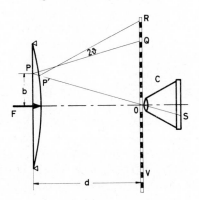

FIG. 6.11. Schematic representation of the apparatus used in Ligtenberg's photoreflective method.

where b/d is the relative field of view of the apparatus depending on the geometry of the plate and its distance from the camera C.

In order to simplify relation (6.15) connecting the fringe distance QR and the angular deflection ϑ, the screen can take a different shape than that of a plane so that the small term b^2/d^2 in eqn. (6.15) becomes zero. Then relation (6.15) becomes

$$QR = 2d \cdot \vartheta. \qquad (6.16)$$

This may be achieved if the screen is a cylindrical surface of a diameter varying between $7d$ and $8d$. By assuming that screen V is a cylinder of a diameter equal to $7d$ and for a relative field of view

$b/d = 0.4$, the error remains smaller than 0.3 per cent. With this shape in the ruled screen the moiré fringes directly yield the partial slope contours of the flexed plate along the principal direction of the rulings. Since slope is not a point function but depends on direction, for the complete evaluation of the slope distribution in the plate it is necessary to obtain a second pattern of partial slopes along another direction. It is convenient to angularly displace the ruled screen through an angle of 90° and to obtain a second moiré pattern of partial slope contours in an orthogonal direction. The method does not yield any means of either numbering the contours or knowing the sign of the slope, but both will be known at some point in the plate from mechanical considerations and this point will be used as a reference point.

If the ruled screen is oriented with its rulings parallel to the axes of symmetry of the plate each of the moiré patterns yields the partial slope distribution along one of the axes. Along any other traverse of the specimen both contour patterns contribute to the evaluation of the slope distribution along the traverse, and the partial slopes in a direction parallel t or normal n to the traverse forming an angle α with the y-axis are given by

$$\left. \begin{aligned} \frac{\partial w}{\partial n} &= \frac{\partial w}{\partial y} \cos \alpha + \frac{\partial w}{\partial x} \sin \alpha, \\ \frac{\partial w}{\partial t} &= \frac{\partial w}{\partial x} \cos \alpha - \frac{\partial w}{\partial y} \sin \alpha. \end{aligned} \right\} \tag{6.17}$$

Figure 6.12a shows the partial slope distribution along a principal axis in a simply supported equilateral triangular plate subjected to uniform pressure, while Fig. 6.12b shows the partial slope contours along the two orthogonal directions of a rectangular perforated plate simply supported along its long edges and free along the short edges, subjected to a load applied at point A.

In order to calculate the magnitude of the moments M_{xx}, M_{yy} and M_{xy} it is necessary to evaluate curvatures $\partial^2 w / \partial x^2$ and $\partial^2 w / \partial y^2$ and twist $\partial^2 w / \partial x \partial y$ from which the moments are obtained via relations

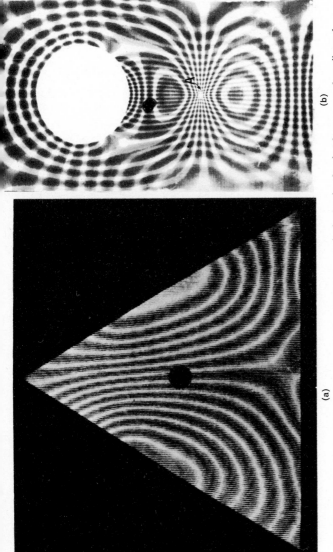

(a)

(b)

Fig. 6.12. Moiré patterns of partial slope contours. (a) Along a principal axis of a simply supported equilateral triangular plate subjected to uniform pressure. (b) Along the two orthogonal directions of a rectangular perforated plate simply supported along its long edges and free along the short edges, subjected to a load applied at point A. *(Courtesy of Dr. F. K. Ligtenberg.)*

$$M_{xx} = -D\left(\frac{\partial^2 w}{\partial x^2} + v\frac{\partial^2 w}{\partial y^2}\right),$$

$$M_{yy} = -D\left(\frac{\partial^2 w}{\partial y^2} + v\frac{\partial^2 w}{\partial x^2}\right), \qquad (6.18)$$

$$M_{xy} = -D(1-v)\frac{\partial^2 w}{\partial x\partial y},$$

where D is the flexural rigidity of the plate and v is Poisson's ratio of the material. The rigidity D is given by

$$D = \frac{Eh^3}{12(1-v^2)}, \qquad (6.19)$$

where E is the modulus of elasticity of the material and h is the thickness of the plate.

Curvatures $\partial^2 w/\partial x^2$, $\partial^2 w/\partial y^2$ and twist $\partial^2 w/\partial x\partial y$ may be readily found from the contour curves of slopes. This can be achieved by a graphical differentiation of the slope distribution curve along any traverse of the plate by displacing this curve by a constant length Δl and measuring the differences in coordinates between the two curves. Practically everywhere the approximation

$$f\left(l+\frac{\Delta l}{2}\right) - f\left(l-\frac{\Delta l}{2}\right)$$

for the first derivative which may be derived from displacing the slope distribution curve $f(l)$ is of sufficient accuracy.

A potential method for obtaining curvature and twist contours from the partial slope moiré patterns is the method of large-field mechanical differentiation. This method consists of superposing two identical patterns of moiré fringes and relatively displacing them through a constant displacement Δx. From the nature of the moiré fringes, which represent lines of constant differences, the new moiré fringes formed by the superposition of the two identical fringe patterns will yield the derivatives along these two axes.

Unfortunately, the method is not useful in cases where the

initial moiré pattern contains sparse fringes. In this case very large displacements Δx or Δy are required to yield a second order moiré pattern containing a useful number of fringes.

The method was introduced in applications of moiré patterns by Dantu[30] who gave a second-order moiré pattern from a displacement moiré pattern by displacing through a distance Δl parallel to the principal direction of the initial pattern of the grating. Duncan and Sabin[31] used the same method for obtaining curvatures and twists from the initial partial slope contours in flexed plates. Recently Parks and Durelli[32] introduced a variation of the method by taking a photograph of the specimen grating and superposing on the same negative the distorted image of the grating, when the specimen is loaded, and displacing it by Δx or Δy. The moiré pattern formed by the two superposed photographs of the undistorted and distorted and displaced gratings directly yield the isoentatics of the field. The inconvenience in this method lies in the difficulty of directly photographing dense gratings of a frequency of 1000 lines per inch. The idea of displacing the gratings instead of the moiré patterns has been used by Heise[33] in the case of flexed plates. He used a modified photoreflective method where the loaded grating of the flexed plate was displaced by Δx or Δy before being photographed in superposition to the image of the unloaded grating.

The main advantage of the double exposure photoreflective moiré method is the fact that plates need not be perfectly flat although, of course, a good reflective surface of the plate is required. The double-exposure technique records changes in slope resulting from the application of load so that if any slight surface irregularity is present in both exposures this is subtracted from the final moiré pattern.

Another advantage of Ligtenberg's photoreflective technique,

[30] Dantu, P., Laboratoire Central des Ponts et Chaussées, Paris, Publ. No. 57–6 (1957). See also, Post, D., *Exp. Mech.* **5** (11) 368 (1965).

[31] Duncan, J. P., and Sabin, P. G., *Exp. Mech.* **5** (1) 22 (1965).

[32] Parks, V. J., and Durelli, A. J., *Trans. ASME, J. Appl. Mech.* **33** (4) 901 (1966).

[33] Heise, U., *Exp. Mech.* **7** (1) 47 (1967).

which is valid for all methods directly yielding partial slope contours in flexed plates, is the fact that they necessitate a single differentiation process to yield the curvatures and twists necessary for the evaluation of principal moments and stresses of the flexed plates. On the contrary, techniques yielding deflection contours of plates necessitate a double differentiation to define the same quantities. Although the deflection moiré patterns are more

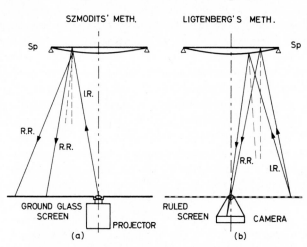

FIG. 6.13. Comparison of principles of moiré formation by Ligtenberg's photoreflective and Szmodits' projection methods.

accurate and denser than the partial slope contour patterns, the double differentiation of the results obtained from these patterns may introduce larger errors than a simple differentiation process.

However, the photoreflective technique presents some limitations and disadvantages. These disadvantages are:

The technique necessitates the use of coarse gratings in order to obtain moiré fringes with good contrast when superposing the unloaded and loaded configurations of the original gratings. Ligtenberg used gratings of pitch $p = 0.0894$ in. Therefore the method necessitates large distances d between the specimen and the grating and relatively large deflections of the plate. Thus different points of the unloaded and loaded plate interfere because of the displacement of the plate during loading (Fig. 6.13b). This source of errors in the method was pointed out by Szmodits,[34] who introduced a variation of the technique which eliminated the errors due to in plane displacement of the loaded plate. Szmodits replaced Ligtenberg's camera placed at a central hole in the ruled screen by a projector projecting a line target on the reflective surface of the specimen. The reflected image of the target was received on a ground-glass plate and photographed by a camera (Fig. 6.13a).

Szmodits' arrangement considerably reduces the transversal displacement of the points in the plate coming to interference, but still necessitates coarse gratings in order to obtain a good contrast and, therefore, a dense moiré pattern can be formed only if the plate is submitted to large deformations, which sometimes may be so large that normal stresses in the middle plane of the plate may create a non-linear relationship between loading and moment distributions, which invalidates the theory of small deflections.

Another disadvantage of the photoreflective method is that the reference ruled screen is approximated to the closest cylindrical surface, which satisfies the simple linear relationship given by eqn. (6.16) between the interfringe spacing and the slope of the plate. This approximation introduces errors which are kept small for certain values of the diameter of the cylindrical screen and distance d between the plate and the screen.

A worthy improvement of Ligtenberg's photoreflective method was introduced by Rieder and Ritter.[35] The changes introduced to the photoreflective method were:

(a) The opaque screen was replaced by a light diffuser D with its principal plane parallel to the longitudinal axis of the optical bench (Fig. 6.14). A denser line grating RG was placed in

[34] Szmodits, K., *Proc. Symp. Shell Struct.*, North Holland Publ. Co., Delft, 1961, p. 208.
[35] Rieder, G., and Ritter, R., *Forsch. Ing.-Wes.* **31** (2) 33 (1965).

contact with the light diffuser and formed a type of multiple source.

(b) On the axis a half mirror M was placed at 45° to the diffuser and was inclined to an angle of 45° to the axis of the bench. The mirror reflected the image of the multisource along the longitudinal axis of the bench zz' and illuminated the reflective surface of the plate P, while it allowed viewing the plate by a camera C placed far away on the zz' axis.

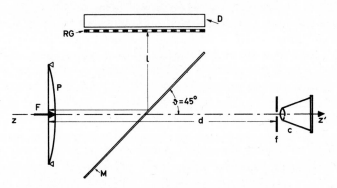

FIG. 6.14. Schematic representation of Rieder's and Ritter's improvements in photoreflective method.

These modifications resulted in the improvements of the technique:

(a) The screen could be denser than the screen used in Ligtenberg's method. While the pitch in Ligtenberg's method was $p = 0.0894$ in., the pitch in Rieder's and Ritter's modification was $p = 0.0202$ in.

(b) The field of view of the camera was considerably reduced by increasing the distance d between the camera and the plate which was now independent of the distance l between diffuser and plate. In the present modification ratio $l/d = 0.129$, while in Ligtenberg's method $l/d = 1$.

(c) The screen surface was flat instead of cylindrical. The error due to the flatness of the multisource grating for the particular dimensions used by Rieder and Ritter was of the order of 0·12 per cent for a maximum relative field of view $b/d = 0·034$, while in Ligtenberg's method the error due to the approximation of the screen to a cylindrical surface with diameter $D = 7d$ was of the order of 0·3 per cent for a maximum relative field of view of $b/d = 0·40$. The 3·4 per cent relative field of view in Rieder's and Ritter's method approximately corresponded to the 40 per cent relative field of view in Ligtenberg's method.

(d) The large values in distance d and the decrease of the pitch resulted in a considerable reduction in the error due to interference of different points in the unloaded and loaded plate.

Finally, the method of Rieder and Ritter was used by Heise[36] with an additional modification in the taking of photographs, which allowed for the direct tracing of partial curvature and twist contours. According to the modification by Heise the image of the screen on the reflective surface of the loaded plate was photographed after displacing the plate along its principal axes x and y by a Δx or Δy displacement respectively. The moiré patterns formed by the interference of the images of the screen reflected on the unloaded and the shifted loaded plate yielded directly the partial curvature contours. The fringe density in these moiré patterns depends on displacements Δx and Δy. Large displacements yield dense moiré patterns, but introduce errors in the correct position of fringes which, in some cases, may be considerable.

If the pitch of the screen is p and the displacements along x- and y-axes are u and v respectively, the moiré fringes of partial slope contours are expressed as

$$S_x(x,y) = \frac{\partial u(x,y)}{\partial x} = kp, \qquad (6.20)$$

where k is an integer.

[36] Heise, U., *Exp. Mech.* **7** (1) 47 (1967).

Suppose that two small and equal displacements $+ \Delta x/2$ and $- \Delta x/2$ are given to the loaded pattern of the screen.[37] Two moiré patterns M_1 and M_2 can be created from the superposition of the loaded patterns with the unloaded image of the screen.

The equation for M_1 is

$$S_x\left(x - \frac{\Delta x}{2}, y\right) = k_1 p, \qquad (6.21)$$

and the equation for M_2 is

$$S_x\left(x + \frac{\Delta x}{2}, y\right) = k_2 p. \qquad (6.22)$$

Equations (6.21) and (6.22) can be expanded to a truncated Taylor series up to the third-order term

$$S(x, y) - \frac{\Delta x}{2}\frac{\partial s}{\partial x} + \frac{(\Delta x)^2}{8}\frac{\partial^2 s}{\partial x^2} - \frac{(\Delta x)^3}{48}\frac{\partial^3 s}{\partial x^3} = k_1 p, \qquad (6.23)$$

$$S(x, y) + \frac{\Delta x}{2}\frac{\partial s}{\partial x} + \frac{(\Delta x)^2}{8}\frac{\partial^2 s}{\partial x^2} + \frac{(\Delta x)^3}{48}\frac{\partial^3 s}{\partial x^3} = k_2 p. \qquad (6.24)$$

The moiré pattern formed by two symmetric translations $\pm \Delta x/2$ is identical to the pattern created by a sole translation Δx. This moiré pattern, called the second-order moiré pattern, is given by the difference

$$nf = (k_2 - k_1)p = \Delta x\frac{\partial s}{\partial x} + \frac{(\Delta x)^3}{24}\frac{\partial^3 s}{\partial x^3}$$

$$\kappa_x = \frac{\partial s}{\partial x} = \left(\frac{nf}{\Delta x} - \frac{(\Delta x)^2}{24}\frac{\partial^2 \kappa_x}{\partial x^2}\right), \qquad (6.25)$$

where f is the interfringe spacing, n the fringe order, and κ_x the partial curvature along the x-axis.

[37] Theocaris, P. S., Discussion on paper by Heise, U., *Exp. Mech.* **8** (8) (1968).

The accuracy of the displacing method depends on the corrective term $\frac{(\Delta x)^2}{24} \frac{\partial^2 \kappa_x}{\partial x^2}$ in eqn. (6.25). If this term is negligible the approximation of representing the partial curvature contours by the moiré patterns formed by displacing the deformed image of the screen by $\pm \Delta x/2$ is satisfactory. When the corrective term is not negligible the following correction procedure may be adopted: If the $\kappa_x = f(x)$ is wanted the approximate $\kappa_x^a = f(x)$ is first traced by neglecting the term $\frac{(\Delta x)^2}{24} \frac{\partial^2 \kappa_x}{\partial x^2}$. Then, the derivative $\frac{\partial^2 \kappa_x}{\partial x^2}$ may be evaluated by a double graphical differentiation of the $\kappa_x^a = f(x)$. This derivative multiplied by the constant factor $\frac{(\Delta x)^2}{24}$ yields a corrective term which, subtracted from the first approximation curve, gives the correct values of curvature. The same procedure can be repeated for higher terms and applied for the determination of the curvature along the other axis and the twist of the plate.

Finally, the possibility of utilizing the photoreflective moiré method to two-dimensional stress problems by applying the slab analogy should be mentioned.

Wieghardt[38] employed plate curvatures of a suitably deformed thin plate in order to determine Airy's stress function in the associated plane stress problem. A thorough discussion of the theory of the slab analogy for simply and multiply connected systems was presented by Mindlin and Salvadori,[39] but it is beyond the scope of this book to discuss the possibilities of this analogy.

The principle of the slab analogy method may be summarized as follows. The solution of a plane stress problem described by Airy stress function F is given by the stress components expressed as

[38] Wieghardt, K., *Forsch. Ing.-Wes.* **49**, 15 (1908).
[39] Mindlin, R. D., and Salvadori, M. G., *Handbook of Experimental Stress Analysis*, M. Hetényi (Ed.), Wiley, New York, 1950, ch. 16.

$$\left.\begin{aligned}
\sigma_x &= \frac{\partial^2 F}{\partial y^2} + V, \\[2mm]
\sigma_y &= \frac{\partial^2 F}{\partial x^2} + V, \\[2mm]
\tau_{xy} &= -\frac{\partial^2 F}{\partial x \partial y},
\end{aligned}\right\} \tag{6.26}$$

where V is the body force potential.

By setting $F = kw$, where k is a proportionality constant and w the deflection of the plate, the second derivatives in eqns. (6.26) can be expressed by the second derivatives of the deformed plate, which constitutes the analogous system. Based upon these considerations, the state of stress in the two-dimensional problem can be expressed as

$$\left.\begin{aligned}
\sigma_x &= k\frac{\partial^2 w}{\partial y^2} + V, \\[2mm]
\sigma_y &= k\frac{\partial^2 w}{\partial x^2} + V, \\[2mm]
\tau_{xy} &= -k\frac{\partial^2 w}{\partial x \partial y}.
\end{aligned}\right\} \tag{6.27}$$

The main difficulty in applying the photoreflective moiré method to the measurement of partial curvatures and twists of the field necessary for the slab analogy lies in achieving the correct boundary conditions and especially in cases where a simultaneous application of different boundary conditions is needed. In complex load conditions a superposition of test results from different specimens with different boundary conditions is practicable.

The accuracy of the method lies within a few per cent for simple problems of elasticity and almost matches the accuracy of the photoelastic methods. The main advantage of the method over the photoelastic method is that the slab analogy method directly yields the values of stresses without necessitating the elaborate methods of separating stresses as in photoelasticity. The method of slab analogy was extended to the general case of variable elastic or thermal problems by Zienkiewicz and Cruz,[40] who discussed cases of discontinuous variation of temperature and shrinkage. It was shown that when the displacements of the slab are induced by appropriate loads and not, as in previous applications, by imposed edge displacements the accuracy of the method can be considerably improved.

An improvement to Ligtenberg's photoreflective method was recently introduced by De Haas and Loof.[41] The ruled screen used by these investigators had a frequency of the order of 500 lines per inch. This reduction in pitch resulted in a proportional increase of the number of fringes in the moiré pattern exhibiting the slope contours of the plate. Moreover, the photographic negative of the moiré pattern presented fine background lines, the frequency of which was increased to 500 lines per inch. This background grating copying the rulings of the screen was capable of diffracting the light passing through. If the negative is viewed under suitable illumination and in the direction of the first-order diffraction, from these lines a filtering effect is accomplished and the field in between the moiré fringes appears clear of these lines, while the moiré fringes remain dark. This results in much sharper contrast in the moiré pattern than of the pattern obtained with the conventional method.

6.4. The photoreflective moiré method for the evaluation of principal moment and stress directions

Besides the applications of the photoreflective moiré method to the study of flexed plates and two-dimensional stress problems, a graphical technique may be related to this method which allows for the determination of the directions of principal moments and stresses. This technique was introduced by Bouwkamp[42] and it is called the *isoclinic method*.

[40] Zienkiewicz, O. C., and Cruz, C., *Int. J. Mech. Sci.* **4** 285 (1962).

[41] De Haas, H. M., and Loof, H. W., *Exple. Spannungsanalyse*, VDI-Berichte, No. 102, 65 (1966).

[42] Bouwkamp, J. G., *Exp. Mech.* **4** (5) 121 (1964).

The method is based on the fact that moments and stresses are directly related to partial curvatures of flexed plates. Therefore, the directions of principal moments in the flexed plate or the principal stresses in a plane stressed specimen should coincide with the directions of principal curvatures in the flexed plate.

The complete identity between the principal moment and stress directions, as well as the principal curvature directions in the

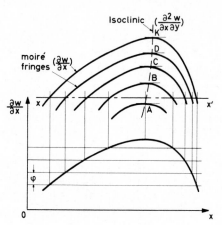

Fig. 6.15. Determination of an isoclinic from moiré fringes of partial slope contours.

flexed plate, allows the application of the graphical method for the evaluation either of the principal moment directions in flexed plates or of the principal stress directions in two-dimensional problems.

The isoclinic method is based on the identification of certain locations of the plate where the principal curvatures coincide with the orientation of the rulings of the grating. At these locations the twist $\frac{\partial^2 w}{\partial x \partial y}$ must be equal to zero.

From Fig. 6.15 it can be shown that for any point on the plate where the rulings of the screen are tangent to the moiré fringes of partial slope contours, the twist must be zero. Hence the principal curvature directions at these points coincide with the orientation of the rulings of the screen. Thus it is possible to determine on each moiré pattern of partial slope contours a sequence of points $(A, B, C, \ldots, K)$ along which the twist becomes zero. This line

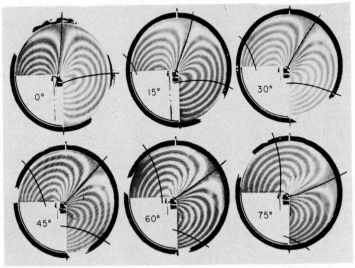

FIG. 6.16. Family of isoclinics for different grating orientations of a circular plate laterally loaded by a concentrated force applied at its centre. The plate is simply supported at its circular rim, a quadrant of the plate is removed and its radial edges are free. (*Courtesy of Prof. J. G. Bouwkamp.*)

actually represents an isoclinic line along which the principal curvature directions coincide to the principal direction of the rulings of the screen. By rotating the ruled screen by a constant angle increment it is possible to obtain a set of isoclinics in exactly the same manner as the photoelastic isoclinics are obtained in a plane polariscope. Figure 6.16 shows the isoclinics for different angles of orientation in a circular clamped plate laterally loaded by a concentrated force applied to the centre of the plate. A quadrant of the plate was removed and its radial edges were free from any support. The isoclinics were traced for each picture by connecting all the points where the tangents to the moiré fringes were parallel to rulings of the screen.

The method is similar to the already described method of direct tracing of isoentatics (§ 2.5). In this method, instead of angularly displacing the reference grating, as in the case of the isoclinic method, the pitch of the grating was slightly changed and the curves along which the tangents to the moiré fringes were parallel to the Cartesian axes of coordinates gave the isoentatics along which the strain component was equal to the increase between the pitches in the reference and specimen gratings.

A graphical method, similar to the method used in photoelasticity for the tracing of principal stress trajectories, may be used for tracing the principal stress or moment trajectories from a pattern of isoclinics obtained from a set of individual isoclinic photographs taken at different ruling orientations.

A pointwise method was developed by the same author for the determination of the principal curvature directions in flexed plates by using a set of two photographs yielding the partial slope contours in orthogonal directions. This method is based on a Mohr-circle procedure and is therefore beyond the scope of this book.

6.5. The Salet–Ikeda moiré method for the determination of partial slopes in flexed plates

The Salet–Ikeda method is an optical method for recording the partial slope contours in flexed plates. The method is similar to the Schlieren technique utilized for recording shock waves and thermal gradients in fluids. The method was introduced in France by Salet[43] and was then modified in Japan by Ikeda.[44] It was used for the measurement of slopes in soap films and other curved surfaces. Duncan and Brown[45] adapted the Salet–Ikeda method for the measurement of partial slopes in flexed plates.

[43] Salet, G., *Bull. Assn. Tech. Mar. Aeronaut.* **43**, 107 (1939).

[44] Ikeda, K., *Proc. First Japan Nat. Cong. Appl. Mech.* 1951, p. 219.

[45] Duncan, J. P., and Brown, C. J. E., *Proc. First Int. Cong. Exp. Mech.*, Pergamon Press, 1963, p. 149.

Essentially the method utilizes a coarse multiple source illuminating the reflective surface of the plate and a slit or pinhole placed at the viewing position of the instrument.

The multiple source consists of a white light diffuser LS and a coarse target T (Fig. 6.17). The targets are reduced photographic negatives of black-and-white line drawings. A variety of targets may be used in the experiments. These may be either parallel line coarse gratings or equispaced circular and radial gratings or dots on a grid which represent a superposition of two orthogonally

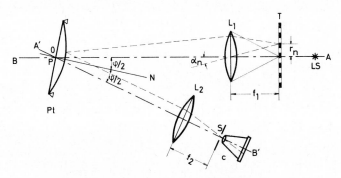

FIG. 6.17. Schematic representation of Salet–Ikeda method
for recording partial slopes in flexed plates.

crossed parallel line gratings. The parallel line gratings are convenient for the formation of partial slope contours in a direction parallel to the principal direction of the rulings, while the circular and radial gratings may be used in cases of polar symmetry. The frequency of these targets entirely depends on the geometry of the specimen, the rate of change of slope of the plate surface and on the degree of accuracy required. From the point of view of accuracy it is desirable to resolve slope as finely as possible whilst ensuring that photographic resolution is not made impossible by an overcrowding of fringes.

A collimating lens L_1 is placed at a distance f_1 from the target T, where f_1 is the focal length of the lens. The parallel light beam after being reflected from the surface of the plate specimen Pt passes through a second convex lens L_2 at the focal distance f_2 of which either a slit (for the case of parallel line targets) or a pinhole (for the case of radial or circular gratings) is placed. The slit S must be parallel to the direction of rulings in the case of parallel line gratings.

If the reflective surface of the plate remains undeformed and plane, the target T will be viewed at the slit, via reflection on the undeformed plate. If the plate is deformed, the image of the target, viewed at the slit, will be distorted.

Consider a circular target and a circle of radius r_n of this target.

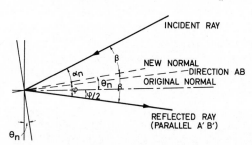

FIG. 6.18. Slope change versus target radius relationships.

This circle will be projected on the reflective surface of the plate Pt and only its reflection, which remains parallel to the axis $A'B'$, is observed at the pinhole. If α_n is the viewing angle of circle r_n it is valid that

$$r_n = f_1 \tan \alpha_n. \qquad (6.28)$$

For the rays to be reflected parallel to the axis $A'B'$ and converge to the pinhole, a simplified analysis based on Fig. 6.18 shows that $\alpha_n = 2\Theta_n$ and therefore

$$r_n = f_1 \tan 2\Theta_n, \qquad (6.29)$$

where Θ_n is the angle of rotation of the surface element of the plate due to loading and α_n is the angle of incidence of the ray impinging on this element. To an observer at the pinhole or the slit many contours of equal slope corresponding to the circle r_n or the line l_n in the target will appear on the specimen plate surface. If a camera is placed behind the pinhole or slit, it may record the pattern of partial slope contours of the plate.

Since the method provides an absolute measure of slope, surfaces of the plates require a careful preparation before testing. However, if the flatness condition of the plate is not satisfactory an initial photograph may yield the elevations from flatness of the plate, which must be subtracted from the loaded pattern of the plate.

The method is in reality an optical method of the Schlieren type and strictly speaking it cannot be classed as a moiré method. But, if the slit or pinhole is considered as a grating with one ruling, this grating may be considered as interfering with the target grating and the method in this case can be classified as a moiré method.

From simple trigonometric considerations it can be shown that a displacement of the slit or pinhole equal to the pitch of the target corresponds to a translation by one fringe of moiré pattern without alteration.

This is of interest when considering the most suitable size of the slit or pinhole. A large slit or pinhole will cover a large apparent range of slope. Because of this a fine target requires that the slit or pinhole be moved a small amount to enable this change to be observed. Moreover, a large slit or pinhole which in itself contains this movement in its width or diameter will lead to a poor picture. A coarse target, however, projected on the same plate specimen which is deformed in such manner as to yield an equal number of slope contours, will require a larger displacement of the slit or pinhole to make one moiré fringe occupy its neighbour's position and hence the size of slit or pinhole may be larger. Generally, coarser targets yield better definitions since the slits or pinholes must always be of finite size.

From the above argument it can be deduced that the size of slit or pinhole solely depends on the pitch of the target and not on the curvature of the specimen.

It was also shown by experience and by this argument that clearly defined slope contours are observed for a slit width or pinhole diameter equal to half the pitch of the target grating.

Since the lenses utilized in the experimental arrangement must be lenses of large focal length, of good quality and of large aperture for experimental work at a reasonable scale, they can be replaced by cheaper front aluminized spherical mirrors, which may further improve the sensitivity of the method through the use of longer focal lengths.

Figure 6.19 shows the moiré pattern of partial slope contours in

FIG. 6.19. Moiré pattern of partial slope contours in a triangular plate centrally loaded and supported at the apices (the target used was a circular coarse grating).

a triangular plate centrally loaded and supported at the apices. The target used was a circular target.

6.6. The reflected image moiré method for recording slope contours in flexed plates

The reflected image moiré method is based on the interference of the reflected image of a line grating projected on the polished front surface of the plate by a parallel light-beam with a second similar grating. The method was introduced by Theocaris[46] and it

[46] Theocaris, P. S., *Proc. Second SESA Intn. Cong. Exp. Mech.*, B. E. Rossi (Ed.), 1966, p. 56; *Exp. Mech.* **6** (4) 212 (1966).

yields a mapping of the partial slope contours of the flexed plate.

The principle of the method is depicted in Fig. 6.20a. A collimated beam of monochromatic mercury green light ($\lambda = 5460$ Å) is incident on the polished surface of the plate, at an angle ϑ with the Oyz plane. It passes, before impinging on the plate, through a reference line grating RG_1 of a frequency of 200 lines per inch, which is placed parallel to the reflecting surface of the plate and at

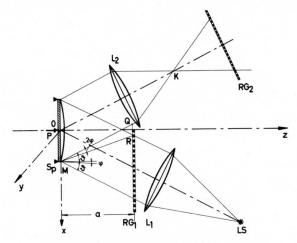

FIG. 6.20a. Schematic representation of the reflected image moiré method for recording partial slopes in flexed plates.

a distance a. The lines of the grating run perpendicular to the Oxz plane. The distance a is selected such that the reflected beam from the plate does not interfere with the reference grating for the particular angle of incidence, which in practice was found to have an optimum value of the order of 10°. The reflected beam from the plate interferes with another grating RG_2, similar to the first grating and having its rulings parallel to the rulings of RG_1.

The apparatus required for the application of the technique is simple. Besides the monochromatic light source LS, the lens L_1 and the reference grating RG_1, the apparatus contains a camera, on the ground-glass screen of which the reflected image of the grating RG_1 is received. The distance between the camera lens L_2 and the ground-glass screen is adjusted to yield an image of the reference grating at unit magnification. This image interferes with the second grating RG_2, which is juxtaposed to the ground-glass screen. The correct position of the lens L_2 and the ground-glass screen can be found by minimizing the number of moiré fringes appearing in the image.

The camera lens L_2 sees the reference grating RG_1 through reflection on the polished surface of the plate. With the plate undeformed and RG_1 parallel to the surface of the plate, the interference of the image of RG_1 and the RG_2 yields the irregularities of the surface of the plate. With the plate flexed the parallel incident rays will deflect through an angle $(\vartheta \pm 2\varphi)$ in the Oxz plane because the surface of the plate at a generic point M is no longer parallel to the reference grating and it forms an angle φ with it.

Consider the case where two points A and B of the plate are rotated through angles φ_1 and φ_2 when the plate is flexed. The camera lens L_2 sees through these points the points C and D of the reference grating respectively before the plate was flexed. After bending of the plate, points C and D are displaced to points C' and D' respectively (Fig. 6.20b). If these points correspond to two successive moiré fringes the difference $(CD - C'D') = p/2$, where p is the pitch of RG_1. From the triangles EFG and GFH, where angles $KEF = \vartheta$, $EGF \approx 90°$ and $FEH = 2(\varphi_2 - \varphi_1)$ and the distance $FH = p/2$, it can be readily deduced that

$$(\varphi_2 - \varphi_1) = \frac{p \cos^2 \vartheta}{2a}.$$

If the number of fringes between arbitrary points of the plate is N corresponding to a variation φ of the angle ϑ then it is valid that

$$\varphi = \frac{Np \cos^2 \vartheta}{2a}. \tag{6.30}$$

The fringe order N of the moiré pattern is proportional to angle

φ, which in turn is proportional to slope $\partial w/\partial n$ in the n-direction, normal to the direction of the rulings. Then the following relation holds between N and w:

$$N = k\frac{\partial w}{\partial n}, \qquad (6.31)$$

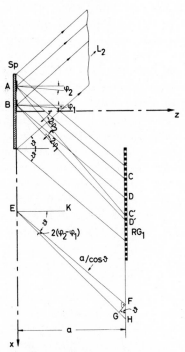

FIG. 6.20b. Geometry of formation of moiré fringes in the reflected-image moiré method.

where k is a factor of proportionality, which depends on the characteristics of the optical arrangement and the frequency of the gratings.

Hence the moiré pattern formed from a grating having its rulings parallel to one axis (say Ox) yields the partial-slope contours of the deflection along an orthogonal direction Oy. Angular displacement of the grating about the Oz axis through an angle of 90° yields another moiré pattern of the partial-slope contours of deflections along the Ox axis.

If the position of the camera lens L_2 is selected so that the interference of the image of RG_1 and RG_2, with the plate unloaded, yields a moiré pattern of minimum moiré density, the moiré pattern, with the plate flexed, directly yields the partial slope distribution of the plate due solely to the application of the load (method A). If glass or Perspex is used as material for the plate, the original moiré pattern depicting the irregularities of the surface of the plate shows only fractions of a fringe because of the satisfactory initial flatness of the plates of these materials. The position of the camera lens L_2 for which the interference of the image of RG_1 and RG_2 yields a minimum of moiré fringes with the plate unloaded is called zero fringe position.

If the grating RG_2 is displaced from the zero fringe position, a number of equidistant fringes parallel to the lines of the gratings will appear which must be algebraically subtracted from the loaded pattern (method B). The introduction of such a pattern facilitates the determination of the slope distribution in cases and areas where the partial-slope distribution along a particular direction varies slowly.

A second camera photographed the moiré pattern formed on the ground-glass screen of the first camera. The photographs before and after loading yielded the original and loaded patterns which, after subtraction, yield the partial slope distribution of the deflection of the plate along a direction normal to the rulings of the gratings. Angular displacement of the gratings through 90° yields another pattern of the partial-slope distribution along an orthogonal direction.

Either ordinary glass or Perspex plates with their front surface made reflecting were used as specimens, which, when flexed, gave very dense moiré patterns.

Along any other traverse of the specimen, both contour patterns contribute to the evaluation of the slope distribution along the traverse, and the partial slopes in a direction parallel or normal to a traverse forming an angle α with y-axis are given by eqns. (6.17).

Figure 6.21 presents the partial slope contour pattern along the principal direction of a square plate supported at the four corners and loaded by a concentrated load at the centre of the plate.

Figure 6.21a shows the contour pattern of partial slope in the case where RG_2 is at the image of RG_1 and no initial pattern appears with the plate unloaded. Figures 6.21b, c show the moiré patterns with the plate loaded under the same conditions as Fig. 6.21a but with RG_2 displaced forward or backward from the image of RG_1. While the same partial-slope contours can be derived from all three patterns, with the latter two a detailed mapping is easier in areas where the first pattern shows sparse fringes.

Indeed, a displacement of lens L_2 in either side of its zero fringe position corresponds to an introduction of a fictitious constant slope all over the plate. The moiré fringes formed in this case are parallel equidistant fringes as those shown in Figs. 6.21b, c. This fictitious constant slope is added algebraically to the variable slope distribution due to loading. Therefore if this initial constant slope is subtracted from the total slope distribution in the plate, it yields the slope distribution due to loading. If the initial slope distributions shown in Figs. 6.21a–c are algebraically subtracted from the corresponding slope distributions along any transverse of the plate, they yield an identical slope distribution.

The effect is analogous to the effect obtained in the linear differential moiré method (§ 2.4) where linear disparities in pitch in the reference grating are used to introduce a fictitious strain field in the moiré pattern of the specimen, which has the advantage of increasing the number of fringes in the moiré pattern thus facilitating the evaluation of the displacement distribution in the strain field.

The same effect could be obtained if the lens L_2 remained at its zero-fringe position and the RG_1 was rotated about a y-axis. However, in this case the distance a will not remain constant, and therefore the accuracy and the simplicity of the technique impaired.

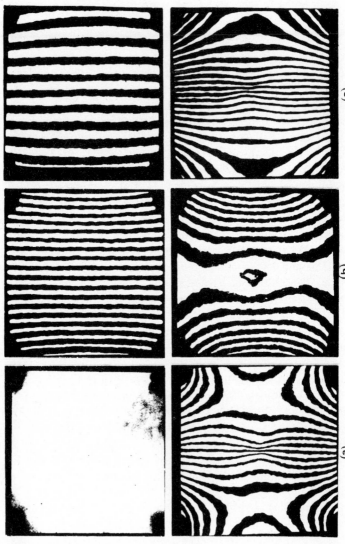

FIG. 6.21. Moiré patterns of partial slope contours along the principal directions of a square plate

In both cases the actual partial slope diagram along any section can be derived from subtraction of the unloaded diagrams along the two orthogonal directions of orientation of the gratings from their corresponding loaded diagrams.

Differentiation (graphical or numerical) of the partial-slope distribution along any direction yields the variation of curvature along the same line. Determination of the partial curvatures $\partial^2 w/\partial x^2$ and $\partial^2 w/\partial y^2$ and the twist $\partial^2 w/\partial x \partial y$ along a grid of orthogonal lines suffices for the evaluation of the moment and strain distribution all over a plate loaded in the elastic range, provided that the flexural rigidity of the plate has been determined.

The fringe order of the resultant moiré pattern, after subtraction of the unloaded pattern, depends on the optical characteristics of the apparatus and especially on angle ϑ and distance a, as well as on the frequency of the line gratings RG_1 and RG_2. For an absolute calibration a complete evaluation of the slope distribution necessitates the accurate measurement of these quantities. Although the measurement of angle ϑ and the distance a may be easily effectuated and the pitches of gratings RG_1 and RG_2 are known, sometimes it is easier to interrelate the moiré fringes and the true partial slope of the plate by using a relative calibration. In this case a calibration plate may be used to yield the absolute value of the slope for the particular optical characteristics of the apparatus used for the test. As a calibration plate a square plate in anticlastic loading may serve excellently the purpose.

The square plate must be made from the same material as the specimen plate with the same mechanical characteristics and the same thickness. Thus the flexural rigidities of the calibration and the specimen plates are the same.

It is well known that a square plate in anticlastic bending, created by two pairs of oppositely applied normal loads P at its diagonal corners, is in a state of constant pure bending, and the principal bending moments are

$$M_x = -M_y = -\frac{P}{2},\qquad(6.32)$$

where the Cartesian system of coordinates Oxy has its origin at the centre of the plate and its axes coinciding with the diagonals of the plate. The moiré pattern showing the partial slope contours is formed by straight and equidistant fringes parallel to the one diagonal of the plate and spaced at a distance f.

The equation for the deflection w of the plate is given by[47]

$$w = \frac{P}{4D(1-v)}(x^2 - y^2), \tag{6.33}$$

where D is the flexural rigidity of the plate and v is Poisson's ratio for the material. Introducing the expression for rigidity in the last relation and taking the derivative of the deflection with respect to the x-axis it is found that

$$\frac{\partial w}{\partial x} = \frac{6P(1+v)}{Eh^3}x, \tag{6.34}$$

where E is the modulus of elasticity of the material of the plate and h the thickness of the calibration plate.

By measuring the distance x between two successive moiré fringes (which expresses the interfringe spacing f) and the corresponding load P applied to the anticlastic plate and introducing these values to eqn. (6.34), the absolute value of slope $\partial w/\partial x$ for the calibration plate can be evaluated. Since the mechanical characteristics of the calibration plate and the tested plate are the same and the optical characteristics of the apparatus unchanged during the two experiments, the absolute values for the partial slopes $\partial w/\partial x$, $\partial w/\partial y$ can be determined.

Thus by measuring only the applied load and the interfringe spacing in the calibration plate the apparatus can be easily calibrated.

The same technique may be used in the case where the flexural rigidity of the plate must be evaluated when the optical characteristics of the apparatus are known.

[47] Timoshenko, S., and Woinowsky-Krieger, S., *Theory of Plates and Shells*, 2nd edn., McGraw-Hill, New York, 1959, p. 42.

Figure 6.22 presents the moiré pattern of a square Perspex plate loaded in anticlastic bending. Figure 6.23 presents the slope distribution in a square plate supported at the four corners and loaded by a concentrated central load, along traverses parallel to the sides of the plate and separated by a distance equal to one-eighth of the width of the plate. In the same figure the values of the partial slopes

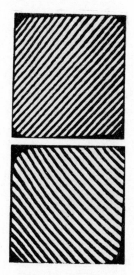

FIG. 6.22. Moiré patterns of partial slope contours of square plates in anticlastic bending.

are given as determined theoretically by the finite difference solution of the bi-harmonic equation for plates given by Marcus.[48]

The coincidence of theoretical and moiré results is satisfactory. This was expected because the moiré patterns precisely represent the partial slope contours of the plates and, on the other hand, the deflection of the plates was kept very small and hence the membrane action of the plates insignificant.

[48] Marcus, H., *Die Theorie elastischer Gewebe und ihre Anwendung auf die Berechnung biegsamer Platten*, 2nd edn., J. Springer, Berlin, 1932, ch. 8.

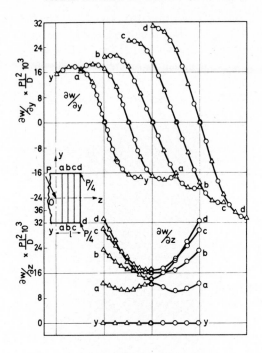

FIG. 6.23. Comparison of theoretical (circles) and experimental values (triangles) of slope distribution along traverses parallel to the sides of the plate and separated by a distance equal to one-eighth of the width of the plate. The square plate is supported at its corners and centrally loaded by a concentrated load.

The sensitivity of the method depends on the fineness of the gratings used and on the distance a between the specimen and the reference grating RG_1.

If Δu is the local shift of the image of RG_1 due to the reflection on the deformed plate Sp through an angle φ, this shift is expressed by relation (6.34)

$$\Delta u = \frac{2a\Delta\varphi}{\cos^2\vartheta}. \qquad (6.35)$$

During the particular experiments, distance a was equal to 160 mm (6·3 in.) and $\vartheta = 10°$. The frequency of the reference grating was 200 lines per inch. If a moiré fringe can be located to within one-fifteenth of the interfringe spacing, then

$$\Delta u = \frac{1}{120}\,\text{mm}, \qquad (6.36)$$

$$\Delta\varphi = \frac{\Delta u \cos^2\vartheta}{2a},$$

$$\Delta\varphi = 2·5 \times 10^{-5}\,\text{rad}.$$

The sensitivity of $2·5 \times 10^{-5}$ rad is very satisfactory and small deflections of the plate can be measured by moiré fringes.

The reflected image moiré method is simple, reliable and easily adapted to any material and any case of loading. It may be applied to dynamic problems since it does not necessitate any kind of isolation from external vibrations. It is also suitable for the study of the mechanical properties of viscoelastic materials.

The accuracy of the method is high because the moiré patterns obtained are directly related to the partial slope distribution of the plate, while, in previous methods, the slope contours are only approximated by the moiré fringes, the degree of approximation depending on the geometry of the apparatus.

The quality of the reference gratings influences the sharpness of the moiré fringes. Good reference gratings, having either a transmittance of 50 per cent or complementary transmittances, yield sharp and well-defined moiré fringes all over the field. The frequency of the gratings used in the experiments was of the order of 200 lines per inch. This frequency is rather low and the photographic reproduction of a master yields very satisfactory copies, which in turn give very sharp moiré fringes.

Comparison of the results obtained by the moiré method with

the theoretical results yielded a figure of accuracy of the method of the order of 2–3 per cent. The larger discrepancies were encountered at boundaries.

The sensitivity was of the order of $2·5 \times 10^{-5}$ rad for a frequency of the reference grating of 200 lines per inch, which was considered to be more than sufficient for the case of flexed plates. A result of the high sensitivity of the method was the fact that glass plates can be used as specimens. Although the deflections of these plates are kept very small, the resulting moiré patterns are very dense.

The method is versatile and adequate to measure small slopes as well as large slopes by changing only the frequency of the reference gratings of the plate. It may then be used in problems of small elastic deflections as well as of large plastic deflections.

All these characteristic quantities of the reflected image moiré method make this technique the most adequate and advantageous method for the study of flexed plates from any other similar moiré method.

6.7. The multisource and the slit source and grating methods for recording slope contours in plates

The multisource and the slit source and grating moiré methods can be extended to record partial slope contours of flexed plates. Both methods were applied for the study of flexed plates by Theocaris and Koutsambessis.[49] The methods as used to the study of two-dimensional transparent specimens loaded in plane stress were modified in order to be adapted to the study of flexed plates. In the multisource moiré method the front surface of the flexed plate acts as a mirror and distorts the already formed moiré pattern by the two gratings placed at some distance from each other and in the path of a collimated light beam.

The experimental arrangement used was simple. Two line gratings were mounted in parallel planes and at a distance a apart and at some distance from a monochromatic diffused light source.

[49] Theocaris, P. S., and Koutsambessis, A., *Acta Tech. Czech.* **18** (6) 176 (1967).

The distance a was taken approximately equal to 1 in., while the frequency of the gratings was 200 lines per inch. The rulings of both gratings were oriented either parallel to the x-axis or parallel to the y-axis of the plate specimen. At a distance $b = 2$ in. from the mid-plane of the gratings the plate specimen with its front surface specularly polished was reflecting the obliquely incident light beam

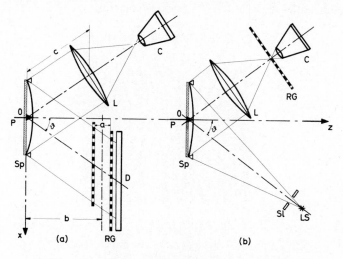

FIG. 6.24. Schematic representation of the optical system used in the multisource and slit-source and grating methods for recording slope contours in flexed plates.

(Fig. 6.24a). A lens L placed at a distance $c = 12$ in. from the plate received the reflected beam and concentrated it to a camera C.

Since the moiré pattern is formed in a plane different to that of the specimen, according to the theory developed in § 5.2, in order to have both specimen and fringe pattern focused on the camera back, the camera must be located at such a distance and the diaphragm stop must be chosen such that the depth of focus of the lens covers the distance between the plate and the fringe pattern planes.

With the plane specimen undeformed and flat, no moiré pattern appears on the ground-glass screen of the camera, i.e. if the optical bench is arranged to correspond to the zero fringe position. If the plate is flexed the moiré pattern formed yields the map of partial slope contours on the principal direction of the gratings.

Figure 6.25 shows the moiré patterns of partial slope contours of a flexed square plate centrally loaded and simply supported. The pattern exactly coincides with the corresponding pattern obtained by the reflected image moiré method.

Fig. 6.25. Moiré pattern of partial slope contours of a flexed square plate centrally loaded and simply supported as obtained by the multisource method.

Similar results can be obtained by placing one grating in close proximity to the reflective surface of the plate (that is up to 0·5 in.). In this case the second grating is replaced by the virtual image of the reference grating, and the moiré fringes thus formed are identical to those of the limiting case of two identical gratings separated by an air gap (see p. 173).

It is possible to replace one grating in the reflected image moiré method by a slit and obtain the same result as it was proved in § 5.3 in the case of two-dimensional transparent specimens in plane stress.

A rectangular slit source, comprised of a mercury monochromatic point source and a rectangular slit of width equal to 1 mm and of a large length compared to its width, formed the illuminating unit. At a large distance from the illuminating unit the plate specimen was placed, on the specularly polished front surface of which the incident light impinged obliquely at an angle ϑ (Fig. 6.24b). The reflected light passed through a positive lens L at the focal distance of which a line grating RG was interposed with the principal direction of its rulings parallel to one of the axes of symmetry of the plate. A camera C viewed the grating from a distance equal to the focal length of the camera lens.

Provided that the lens L was free of aberrations and other defects and that the front surface of the plate was flat, the image of grating was formed undistorted at minus infinity with an infinite pitch. When the grating was approached to or receded from the lens L the undistorted image of RG was formed somewhere between L and minus infinity. A moiré pattern consisting of equidistant and parallel fringes appears all over the field of the plate, which constitutes the original pattern, and it must be algebraically subtracted from the loaded pattern in order to yield the distribution of partial slopes along the plate.

With the plate flexed by the application of an external load, its front surface becomes curved and the reflection of the incident light beam deviates from its initial position according to the variable increase or decrease of the angle of incidence of light. Accordingly, the shape of the moiré pattern was distorted following the distortion of the image of the grating. The moiré fringes formed by this arrangement again represent the partial slope contours of the plate along the principal direction of the rulings of the grating.

As an illustrative example the method was applied to the problem of a centrally loaded and simply supported square plate (Fig. 6.26). The moiré patterns are identical to the moiré patterns obtained by applying the multisource method. While pattern (a) corresponds to zero fringe position, patterns (b) and (c) are derived by displacing RG from the zero fringe position.

As a further example the method was used for the recording of

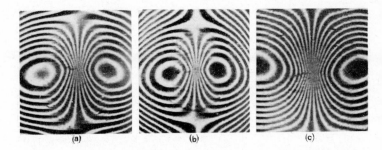

Fig. 6.26. Moiré patterns for the same square plate of Fig. 6.25
obtained by the slit source and grating method.

the moiré fringes of partial slope contours in a triangular plate
subjected to a central load and supported at its three corners. The
moiré pattern of the partial slope contours is shown in Fig. 6.27.

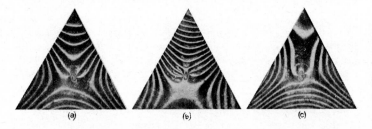

Fig. 6.27. Moiré patterns of partial slope contours in a triangular
plate centrally loaded and supported at its three corners.

The method is very sensitive. It suffices to say that a thin glass
plate 3 in. square yields a very dense moiré pattern by applying a
central load of the order of a few grams.

CHAPTER 7

Applications of Moiré
Methods to Dynamic Problems

MOIRÉ methods are applicable to dynamic problems. These applications may be classed, rather broadly, into three main cases, i.e. the problem of the study of the dynamic response of plates and other structural elements vibrating either freely or submitted to a forced mode of vibration, the problem of stress and strain distribution in two- and three-dimensional bodies submitted to transient loads and, finally, the problem of propagation of shock waves in solids.

In order to show the potentialities of moiré method to dynamic problems, three examples will be treated in this chapter which are related to each of the three distinct problems mentioned above. In all these examples, line or crossed gratings were used as reference gratings, while the specimen gratings were either similar to reference gratings or distorted images of the reference gratings reflected from a reflective surface of the dynamically loaded specimen.

The first illustrative example is concerned with the study of nodal patterns and amplitude distributions on the surface of a freely vibrating plate according to one of its vibration modes. The form and the distribution of vibration nodes and antinodes of the plate may be depicted by two techniques, i.e. either the reflected image moiré method or the dispersion method of a parallel light beam.[1] The amplitude and phase distribution on the vibrating plate may be readily studied by either the photo-

[1] Theocaris, P. S., *Brit. J. Appl. Phys.* **18** (4) 513 (1967); *Exp. Mech.* **8** (5) 237 (1968).

reflective moiré method[2] or by the reflected image moiré method.[3] For the complete analysis of a vibrating plate according to one of its natural modes, only the reflected image moiré method is suitable since it is capable of simultaneously yielding the nodal distribution of the plate as well as its amplitude and phase distribution.

The second example is concerned with the application of moiré method to the study of the transient stress and strain distribution in two-dimensional transparent specimens subjected to impact loading.[4] In this case the linear differential moiré method was used in combination with photoelasticity so as to separate the principal stresses. The method can be used to completely solve the strain distribution problem, but in this case a series of successive tests is necessary to separately evaluate each strain component. Therefore the loading of the specimen or specimens must be of a reproducible nature since the two or three patterns needed must be obtained in independent loadings. Either the contact or the image differential moiré methods are appropriate for this purpose.

The third example is concerned with the application of moiré method to the study of shock-wave propagation phenomena in transparent materials.[5] The linear differential moiré method was used since the deformation field behind the shock wave was not expected to be very large. In this problem Perspex spheres were exploded by explosive charges placed at a central cavity of the spheres. The specimen grating, which was a line grating and occupied a meridian section of the sphere, was imaged at unit magnification on the ground-glass screen of a still camera where the reference grating interfered with the image of the specimen grating to form the moiré pattern of the displacement field due

[2] Nickola, W. E., *Exp. Mech.* **6** (12) 593 (1966), and Theocaris, P. S., *Exp. Mech.* **8** (8) (1968).

[3] Theocaris, P. S., *Brit. J. Appl. Phys.* **18** (4) 513 (1967).

[4] Riley, W. F., and Durelli, A. J., *Trans. ASME J. Appl. Mech.* **29**, 23 (1962).

[5] Theocaris, P. S., Marketos, E., and Gillich, W., *Int. J. Mech. Sci.* **8**, 739 (1966), and Theocaris, P. S., Davids, N., Gillich, W., and Calvit, H. H., *Exp. Mech.* **7** (5) 202 (1967).

to the propagated shock wave. Therefore, the last illustration describes a problem where the linear differential method and the image interference technique were used.

7.1. The moiré methods applied to vibration problems

The experimental study of the vibration of plates started with Chladni's famous experiments executed as early as 1787.[6] The ingenuity of Chladni's experiments depended upon the facts that he employed a bow to excite his plates and used fine sand to obtain the nodal figures of the vibrating plates. It was much later, on the basis of extended experimental evidence, that fine, dry sand was found to be the best powder for sonic frequencies. Chladni, then, has been the originator of the so-called sand patterns rendering visible the various vibration modes of plates.

More recent experimenters used other powders instead of sand to visualize the so-called Chladni figures. Among these powders were Lycopodium spores, having the advantage of giving a fine and uniform powder. Other powders were fine cork, sawdust, flour, stannic oxide, fine silver sand, iodized salt and others. These powders concentrated either to standing areas corresponding to nodal lines or at antinode areas fiercely vibrating. It was Faraday[7] who showed that the occurrence of air current above the surface of the vibrating plate was responsible for the dust clouds which were degenerated afterwards into antinodal figures of the usual nodal figures.

Waller, in 1932, introduced the solid carbon dioxide as an exciter of vibrations in plates. She investigated exhaustively and very adequately the free vibrations of circular and square plates. Her numerous publications have been included in a posthumous book which appeared in 1961.[8]

While the experimental study of flexural modes for plates has

---

[6] Chladni, E. F. F., *Entdeckungen über die Theorie des Klanges*, Breitkopf & Härtel Publ., Leipzig, 1787.

[7] Faraday, M., *Phil. Trans. Roy. Soc. London* **121**, 299 (1831).

[8] Waller, M. D., *Chladni Figures, A Study in Symmetry*, Bell, 1961.

received a rather restricted attention, the study of quartz oscillations by Dye[9] using two-beam interferometry, and mainly by Tolansky,[10] using multiple interferometry, has been rather extensive and complete.

7.1.1. *The dispersion method of parallel light beam*

The principle of the method is based upon the dispersion of reflected rays of light from a vibrating surface. This is not a moiré

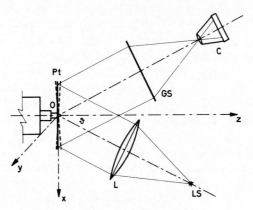

FIG. 7.1. Schematic diagram of the apparatus used in the dispersion method of parallel light beam.

method, but it is included in this section as an introduction to the reflected image moiré method and of its illustrative example.

The simple optical method developed in this section[11] is capable of determining the natural frequencies as well as the nodes and antinodes of plates submitted in flexural vibration. The method is based on the intensity displacement to the nodal areas of a parallel light beam reflected from the polished surface of the plate due to the wavy form of the vibrating plate.

[9] Dye, W. D., *Proc. Roy. Soc. London* **A138**, 1 (1932).

[10] Tolansky, S., *Surface Microtopography*, Longmans, London, 1960, p. 167.

[11] Theocaris, P. S., *Exp. Mech.* **8** (5) 237 (1968).

The principle of the method is shown in Fig. 7.1. The figure exhibits a section of the plate *Pt* by a plane *Oxz* normal to the plane of the plate and a parallel beam of monochromatic mercury-green light, which is obliquely incident at an angle ϑ to the polished surface of the plate. With constant intensity of the incident beam the reflected beam will appear with intensity displacements due to eventual wavy structure of the specimen. If the plate always remains with the same orientation, the reflected beam impinging on the ground-glass screen *GS* will have continuously the same intensity. On the contrary, other parts of the plate whose

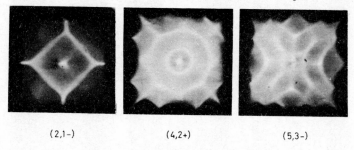

(2,1−)   (4,2+)   (5,3−)

FIG. 7.2. Nodal patterns of a square plate vibrating freely from its centre at its (2,1−), (4,2+) and (5,3−) natural modes.

orientation is changing during deformation will cause the reflected bundles on their surface to deviate from their initial position according to an angle 2φ if φ is the angular displacement of an arbitrary area of the plate. The deviated reflected bundles will appear as darker areas on screen *GS*. The intensity of illumination of each area of the image of the plate on the screen *GS* depends on the angle of rotation φ of each element of the plate during loading. This geometric contrast is on the basis of separation of nodal from antinodal areas of a freely vibrating plate. In fact, the nodal regions remain at relative rest, the only component of displacement being normal to the surface and insignificant. Therefore they will appear on the screen *GS* as bright areas. On the contrary, antinodal regions will vibrate fiercely in various ways and the reflected rays from these regions will be scattered in various directions and outside the plate. This scatter of light will result in a darkening of these regions. Figure 7.2 shows three photographs of the nodal patterns of a square plate vibrating at its $(2,1-)$, $(4,2+)$ and $(5,3-)$ natural modes. All photographs are taken with a completely open diaphragm at the camera $(f = 32)$.

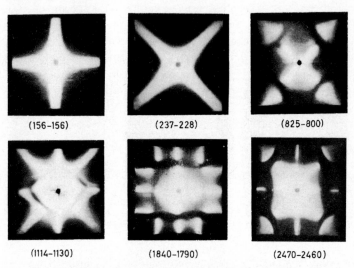

(156–156)   (237–228)   (825–800)

(1114–1130)   (1840–1790)   (2470–2460)

FIG. 7.3. Nodal patterns of a square plate vibrating freely from its centre at its $(1,1)$, $(2,0-)$, $(4,0-)$, $(3,1-)$, $(3,3)$ and $(5,1-)$ natural modes.

The plates tested were made from a sufficiently flat brass alloy sheet and were polished on one surface. The plates were cut square of a side $b = 100$ mm (4 in.) and a thickness $t = 0.74$ mm. They were supported at their geometric centre from the stem of the shaker through a small screw and washer, and they were free on all edges. The centre of the plate follows the forced vibration of the shaker, which is maintained constant at each step of vibration and of limited amplitude in comparison with the amplitudes at antinodes at resonant frequencies. With this arrangement these modes of vibration can only be excited for which the centre of the plate is nodal. In order to obtain the modes with antinodal centres the point of support of the plate must be displaced by an appropriate jig to a pair of symmetric neighbour nodal points.

Figure 7.3 presents six characteristic vibration modes of the square plate. The pairs of numbers underneath each nodal configuration represent the theoretical values of natural frequencies of each mode according to Ritz's method[12] and the experimental values. The modes exhibited in the figure are the $(1,1)$, $(2,0-)$, $(4,0-)$, $(3,1-)$, $(3,3)$ and $(5,1-)$. For the calculation of theoretical values of frequency for the above modes, besides the geometric characteristics of the plates, use was made of the mechanical properties of the plate material which were: modulus of elasticity $E = 1 \cdot 055 \times 10^6$ kg/cm^2 $(15 \times 10^6$ lb/in$^2)$ and Poisson's ratio $v = 0 \cdot 333$. The frequency f_i of the i-mode is given by

$$\omega_i = 2\pi f_i = \frac{\alpha_{mn}}{l^2} \sqrt{\frac{D}{\rho t}}, \qquad (7.1)$$

where D, t and l are the flexural rigidity of the plate the thickness and the side of the square and ρ the density of the material. The coefficients α_{mn} may be derived from the Rayleigh–Ritz series equation[12] and were calculated for $E = 1 \cdot 055 \times 10^6$ kg/cm^2, $v = 0 \cdot 333$ and $\rho = 8 \cdot 32 \times 10^{-3}$ gr$^{(*)}$. sec^2/cm^4 to be:

$$\alpha_{(1,1)} = 12 \cdot 90 \qquad \alpha_{(2,0-)} = 19 \cdot 61$$

$$\alpha_{(3,1-)} = 68 \cdot 37 \qquad \alpha_{(4,0-)} = 117 \cdot 91$$

$$\alpha_{(3,3)} = 152 \cdot 22 \quad \text{and} \quad \alpha_{(5,1-)} = 203 \cdot 82$$

The basic frequency $f_{(1,1)}$ was found equal to

$$f_{(1,1)} = 156 \text{ Hz.}$$

[12] Ritz, W., *Annln. Phys., Leipzig* **28**, 737 (1909).

This frequency coincides with the same mode frequency measured by Waller for a brass square plate having approximately the same mechanical properties and dimensions $l = 10$ cm and $t = 0.1$ cm. The first frequency for this plate was $f_{(1, 1)} = 224$ Hz. If it is taken into consideration that the frequency is proportional to the thickness of the plate t and inversely proportional to the square of l, it is found that both results are coincident. Furthermore, it is valid for square brass plates ($v = 0.333$)[13] that

$$f_{(1, 1)} = 0.6277\frac{ct}{l^2}, \qquad (7.2)$$

where c is the velocity of sound in brass plates given by Wood and Smith[14] as $c = 35.6 \times 10^4$ cm/sec. Then, for the dimensions of the plates used $f_{(1, 1)} = 156$ Hz.

Since the vibrating plates are rigidly fixed at their centres while oscillating, this method not only cuts out antinodal modes but also produces nodal configurations which may not represent normal modes. These modes are imperfect vibration figures occurring in plates either showing slight deviations from perfectly symmetric shapes and eventually mechanical anisotropy or supported at a non-symmetric point. These compound vibration modes happen when two of the normal modes have approximately equal frequencies. The optical method introduced in the section gives the possibility to study the evolution of compound modes, which are very unstable since each phase of development of the non-steady mode can be easily recorded. This possibility is excluded when nodal patterns are recorded by any of the powder-using methods.

Figure 7·4 shows the superposition in various phases of the two simple $(2, 1)$ modes (i.e. two horizontal plus one vertical or two vertical plus one horizontal nodes) to form the $(2, 1 \pm)$ mode which appears very unstable. The relative frequencies of these modes are equal and they are related to the basic frequency $f_{(1, 1)}$ for the $(1, 1)$ mode by the relation $f_{(2, 1 \pm)} \approx 2.7 f_{(1, 1)}$. If the amplitudes of the

[13] Waller, M. D., *Chladni Figures, A Study in Symmetry*, Bell, 1961.

[14] Wood, A. B., and Smith F. D., *Proc. Phys. Soc. London* **47**, 149 and 185 (1935).

two simple normal modes are unequal the pattern of the $(2, 1 \pm)$ mode has the shape of Fig. 7.4d.

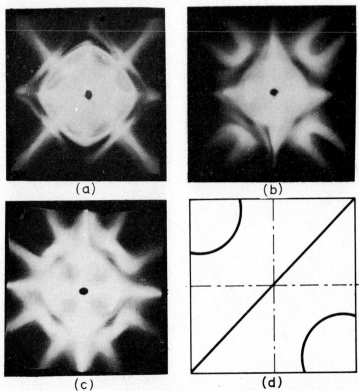

(a)

(b)

(c)

(d)

FIG. 7.4. Nodal patterns of compound unstable $(2, 1 \pm)$ modes of a freely vibrating square plate.

7.1.2. *The reflected image moiré method*

The reflected image moiré method as developed in § 6.6 for the study of the partial slope contours in flexed plates can be used to study the position of nodes and antinodes of plates in flexural vibrations as well as to determine the actual local amplitude and phase distribution.[15]

[15] Theocaris, P. S., *Brit. J. Appl. Phys.* **18** (4) 513 (1967).

The principle of the method is presented in Fig. 6.20a for a statically flexed plate. Fig. 7.5 shows the optical arrangement for a dynamically loaded plate. In this case the static loading jig was replaced by an electromagnetic shaker in order to excite the different vibration modes in the plate. The frequency of the

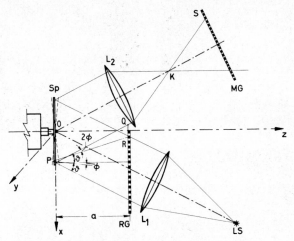

Fig. 7.5. Schematic diagram of the apparatus used in the reflected image moiré method applied to free vibrations of plates.

reference grating was chosen equal to 200 lines per inch and the grating may be either a crossed or a line grating.

If *RG* and the reflective surface of the plate are parallel to each other, the initial moiré pattern depicts the irregularities of the surface of the plate. If *MG* is displaced from its initial position of zero moiré fringe a number of equidistant fringes parallel to the rulings appear. The introduction of such a pattern facilitated the determination of the nodal and antinodal regions of the vibrating plate since it increased the number of moiré fringes along the nodes.

Since for most cases of freely vibrating plates the maximum amplitudes appear along a line of symmetry in the plate (median or diagonal) normal to the corresponding node of vibration, a proper arrangement of gratings RG and SG forms a moiré pattern suitably oriented to yield the total slope distribution of the plate. By integrating this slope distribution the deflection distribution of the plate may be evaluated.

In cases where complicated modes appear it is possible to use crossed gratings for RG and SG and thus simultaneously obtain both moiré patterns of two partial slope contours and evaluate the deflection distribution along any direction of the vibrating plate according to relations (6.17).

While the focal lengths of two lenses L_1 and L_2 influence only the overall dimensions of the apparatus, the sensitivity of the method solely depends on the frequency of gratings, the distance a and the angle ϑ. This can be deduced from eqn. (6.34). By increasing either the distance a or the frequency of the gratings, the minimum angle φ which can be measured is reduced proportionally. However, there is a distance limit a_m for a given frequency of RG beyond which the moiré pattern loses its sharpness. For the particular experimental arrangement, where gratings of 200 lines per inch were used and the parallel light beam was generated by an Osram HBO 200 mercury light point source combined with a convex lens of large focal length ($f = 600$ mm), distance a_m was of the order of 180 mm. This distance may be further increased by using a more intense light source and by arranging the optical system to give a truly parallel light beam.

If a plate is set vibrating in one of its modes, its surface will vibrate in various ways and certain regions will have only a normal component to the surface. The reflected rays from these regions will be imperceptibly displaced while the plate is vibrating and the moiré fringes due to them will remain at rest. The unchanged appearance of the moiré fringes at these regions will reveal the areas of the plate which remain at relative rest during vibration. These are the nodal lines of the plate. On the contrary, regions of plate at the vicinity of antinodes will be set to a rapid oscillatory movement with a considerable angle of rotation about their rest position. In these regions the moiré fringes will vanish entirely.

Then the whole surface will be mapped out into nodal and anti-nodal regions if the plate is illuminated by a continuous mono-chromatic light source.

By setting the oscillator of the plate into action and slowly increasing the frequency of vibration, the plate can be induced into resonance at various frequencies leading to amplified moiré patterns. These patterns being stationary they yield clear pictures of the vibration modes of the plate. However, these pictures do not offer any information concerning the amplitude distribution within the vibrating system.

The various modes of vibration of the plate may be recognized by the increase in number of nodes. The square plate having a fourfold type of symmetry presents six classes of vibration symmetry, i.e. a complete nodal (two medians and two diagonals nodal) ✳ , a complete antinodal ◇ , two medians nodal and the diagonals antinodal + , two diagonals nodal and the medians antinodal ✕ , one median nodal and the other antinodal ✳, and, finally, one diagonal nodal and the other antinodal ✳.

For an illustration and arrangement of vibration pattern symmetry explaining Chladni square figures, see plates 30 and 31 of the exhaustive book by M. D. Waller,[16] which follows the Rayleigh–Ritz theory of vibrating square plates. Diagrams of the square nodal systems are given where the $(m, n \pm)$ nodal systems appear. The m and n coefficients are positive integers which indi-cate the numbers of nodal lines running in the directions of the sides of the vibrating plate. They can be arranged in a square diagram where the m's correspond to columns and the n's represent the rows. Then, all modes of the vibrating plate may be arranged in the order of their appearance. The $m = n$ systems occupy the one diagonal of the square diagram, while the $(m, n-)$ modes are situated to the right and the $(m, n+)$ modes to the left of this diagonal. The six classes of vibration pattern symmetry appear periodically in every second row and every fourth column.

In order to study the distribution of the actual local amplitudes on the surface of the vibrating plate it is necessary to use a periodi-

16 Waller, M. D., *Chladni Figures, A Study in Symmetry*, Bell, 1961.

cally intermittent-light source at the same frequency as the vibration of the plate. A stroboscopic effect is produced and the moiré fringes are seen in their entirety since the illumination occurs only for a single very short fraction of each cycle of the oscillation. It is possible to bias the illuminating system to move to a different part of the cycle and succeed in finding the instant corresponding to the maximum of the deformation of the plate when the moiré fringes appear at their maximum density.

In the technique using continuous light, it is preferable to introduce a dense initial moiré pattern consisting of straight parallel and equidistant fringes because, in this manner, it is possible to obtain a clear distinction between the nodal and antinodal regions. In the technique using intermittent light, where the actual displacement due to vibration and the phase of deformation of the plate must be evaluated it is desirable to bring MG at zero-fringe position. In this manner the initial moiré pattern is eliminated and the loaded pattern formed during vibration corresponds to the actual slope distribution in the plate.

In order to evaluate the sensitivity of the method it must be taken into account that the pitch p of RG and MG as well as the distance a and angle ϑ influence this sensitivity. The limit for the angle ϑ was found to be of the order of 10°. This angle was the maximum angle for which the distortion of the plate remained insignificant. For a grating of a frequency of 200 lines per inch and $\vartheta = 10°$, distance a may be taken as $a = 160$ mm. For these characteristics of the apparatus it can be readily found (§ 6.6) that the smallest deviation in angle $\Delta\varphi$ due to vibration is given by

$$\Delta\varphi = 2 \cdot 5 \times 10^{-5} \text{ rad.} \qquad (7.3)$$

The sensitivity of $2 \cdot 5 \times 10^{-5}$ rad is very satisfactory, and small deflections can be readily measured by this method in the case of intermittent illumination. For the case of continuous illumination where the blurring of initial moiré fringes at antinodes is sought, the sensitivity of the method depends also on the fringe density of the initial moiré pattern and, in any case, is higher than the above figure.

As an illustrative example the method was applied to the study of vibration modes of freely vibrating square plates treated in the earlier section.

The plates tested were made from a brass alloy sheet having a thickness of 0·7 mm and presenting a satisfactory overall flatness not exceeding two to three interference fringes with mercury-green light. The flatness of the plates proved satisfactory since, with a reference grating of 200 lines per inch and an angle of incidence $\vartheta = 10°$, the initial pattern appeared uniform without any fractional fringes.

The plates tested were supported at their centre from the stem of the shaker through a small screw and washer. By fastening the plate at the stem of the shaker through a screw a permanent deflection was produced at the centre of the plate, which engendered moiré fringes when the plate was at rest. It was judged opportune to introduce a dense initial moiré pattern with the plate unloaded by sliding infinitesimally MG along the OK axis. The initial moiré pattern must be subtracted from the corresponding loaded pattern to yield the partial slope pattern due to the deflection of the plate during vibration. This procedure facilitated the determination of the slope distribution along areas where the slope varied slowly.

A Pye–Ling vibration generator (type V50, MK1) was used for the excitation of oscillations of the plate. This type of shaker is an electromechanic force generator operated by an electronic oscillator (Pye–Ling model E 503) and an amplifier (Pye–Ling model 250-VAP). The system can produce controlled vibrations in a useful frequency range between 20 Hz and 10,000 Hz. The force output in this frequency range is 50 lb.

A mercury light spot-type bulb (Osram HBO 200) and a diaphragm was used as a continuous light source. The intermittent light was produced from a General Radio Co. Type 1531-A Strobotac electronic stroboscope combined with a General Radio Co. Type 1536-A photoelectric pick-off powered by a type 1531-P2 flash delay. The photoelectric pick-off produced a voltage pulse whenever its photocell sensed a difference in reflected light. For this purpose a reflective tape was cemented along the thickness of an edge of the plate and the pick-off was aimed parallel to the surface of the standing plate. The positive pulse produced whenever the tape passed by the pick-off was delayed by a time period determined at will by the flash delay and operated the stroboscope in synchronism with the vibrating plate. The flash delay was

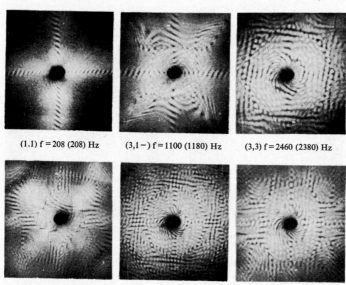

(1,1) f = 208 (208) Hz   (3,1−) f = 1100 (1180) Hz   (3,3) f = 2460 (2380) Hz

(5,1−) f = 3268 (3300) Hz   (6,3+) f = 6250 (6350) Hz   (7,1+) f = 6750 (6850) Hz

FIG. 7.6. Moiré patterns for a square plate vibrating at various natural modes. The first numerals (in parentheses) show the order of mode, while the second group gives the experimental and theoretical values of the natural frequencies.

adjusted to synchronize the flash with the peak of vibration where the maximum deflection of the plate appears. The frozen-in pattern of the deformed plate was then photographed by a still camera. The stroboscope was used for both types of experiments to establish the value of frequency more precisely than the dial reading on the oscillator's console.

Each resonant frequency of the plate corresponding to a mode of vibration was established by defining the maxima in deviations of the moiré fringes. The displacement of the centre of the plate in these frequencies reached either a minimum when the centre was on a nodal line, or a maximum when the centre was an antinode.

Figure 7.6 shows the moiré patterns of various vibrating modes for a square plate supported at its centre. The order of each mode according to the square diagram described previously is given in parentheses, as well as the experimental and theoretical values of natural frequencies for each mode. The theoretical values of

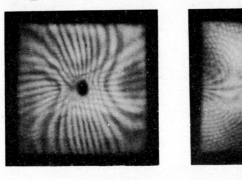
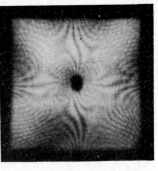

(2,0 –)   (2,1 –)

Fig. 7.7. Moiré patterns of slope distribution in freely vibrating square plates at their (2,0–) and (2,1–) natural modes.

frequencies (given in parentheses) were derived by applying Ritz's method to the square plate.[17]

In the case of the vibrating plate illuminated with a stroboscopic light it was found that the stability of vibration of the plate in its nodal pattern was satisfactory for obtaining standing waves of constant amplitude. Nodal regions remained at relative rest, while antinodal regions were vibrating with a much larger amplitude. By biasing the system to move to a different part of the vibration cycle with the flash delay apparatus, the stroboscopic light source can be adjusted to flash in phase with the maximum of deflection. Thus, not only the amplitude distribution on the surface of the plate can

[17] Ritz, W., *Annln. Phys., Leipzig* **28**, 737 (1909).

be evaluated but also the local phase of the various parts of the vibrating plate can be determined.

Figure 7.7 presents the slope distribution for the $(2,0-)$ and $(2,1-)$ vibration modes of the square plate as it was revealed by the reflected image moiré method with intermittent light. The network of contour curves of equal partial slope along the two princi-

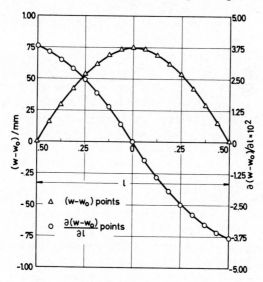

Fig. 7.8. Amplitude and slope distribution along the sides of a square plate freely vibrating at its $(2,0-)$ natural mode.

pal axes of the vibrating plate gives a topographic view of the deformed plate. The hills or troughs of the pattern correspond to antinodal regions and the necks to the nodal lines. Then the three-dimensional configuration of the vibrating plate obtained by an intermittent light illumination yields a complete information of the position of the nodes and antinodes as well as of the local phase and amplitude distribution.

Figure 7.8 presents the amplitude distribution along the sides of a square plate vibrating freely at its $(2,0-)$ mode as derived from the slope distribution of the corresponding moiré pattern at maximum amplitude with the plate at resonance.

The techniques of the reflected image moiré method, where either continuous or intermittent light was used, have proved themselves to be of real value for the study of flexural vibrations in plates. Each of the two methods has its advantages. Continuous illumination yields the exact form and location of the nodes and antinodes of the vibrating plate. Intermittent illumination reveals the actual deformation of the plate.

The method is simple, reliable and versatile, and it can be used for small plates vibrating elastically as well as for large plates vibrating at large vibrations. The method shows much finer detail and precision than previous experimental methods and, besides, it gives the values of the actual local amplitudes and phases of the vibrating plate. This information is far more important in practice.

From comparison of experimental and theoretical results it may be shown that the accuracy of the method is high, i.e. of the order of 2–3 per cent.

7.1.3. *Ligtenberg's photoreflective method extended to vibrations of membranes*

An extension of Ligtenberg's photoreflective moiré method to the study of vibration modes in flexed membranes was introduced by Nickola.[18]

The only differences in the apparatus used by Ligtenberg for recording slopes in statically flexed plates (Fig. 6.11) are that (a) the place of the plate was taken by the vibrating membrane, the surface of which was aluminized in order to become reflective, and (b) the ruled screen was illuminated by two stroboscopic lights which facilitated the taking of photographs of periodic response for the cases of steady-state vibrations.

It was convenient to mount the camera, the ruled screen and the vibrating membrane on a bench so that the optical adjustments could be easily executed. For driving the membrane specimens a 6-in. speaker was used. The speaker was driven by the

[18] Nickola, W. E., *Exp. Mech.* **6** (12) 513 (1966).

amplified output of a signal generator. A pulse generator was used to trigger the stroboscopic light sources at a rate determined by the signal generator. An electric delay unit in the pulse generator yielded the possibility to vary the phase relationship between the stroboscopic flash and the deflected position of the membrane.

For a non-periodic transient loading the light sources were replaced by continuous light sources and the still camera by a Fastax high-speed camera for recording the transient phenomenon of the deflection of the membrane under impact. In this case, and in order to form the moiré pattern of partial slope contours, a first exposure of the film was made by running the entire roll of the film through the high-speed camera with the membrane unloaded. A second exposure of the same film at the same camera speed and with the membrane loaded formed a series of exposures yielding the moiré patterns of partial slope contours of the impacted membrane. The film speed at impact was approximately 5000 frames per second.

It was previously shown (§ 6.3) that the moiré fringes depict the partial slope contours of the flexed membrane in the principal direction of the rulings of the screen. The change of slope $\Delta\varphi$ between successive moiré fringes is related to the distance d between the screen and the membrane and the pitch p of the screen.

Thus
$$\Delta\varphi = \frac{p}{2d} \text{ rad/fringe.}$$

In the case of these experiments $p = 0.1$ in. and distance $d = 25.0$ in.

Figure 7.9 shows the moiré patterns for a circular membrane clamped at its circumference and vibrated by a uniformly applied and periodic loading. The partial slope contours are given for frequencies $f_1 = 817$ Hz and $f_2 = 870$ Hz. The small dark spot in the central position of the photographs is the reflected image of the hole in the ruled screen, which provided a viewing aperture for the still or high-speed cameras. The slope must be measured in a direction normal to the rulings which are visible in the photographs. For the location of the zero fringe order it is convenient to use the boundary condition for the clamped circular membrane, which yields that the slope must be zero in a direction tangent to the boundary.

Comparison between the experimental and the theoretical

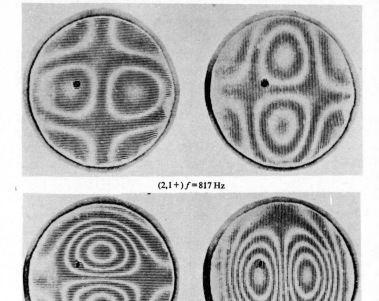

$(2,1+) f = 817$ Hz

$(2,0-) f = 870$ Hz

FIG. 7.9. Moiré patterns of slope contours of a circular membrane clamped at its circumference and vibrated by a uniformly distributed periodic loading (natural modes $(2,1+)$ and $(2,0-)$ are shown). (*Courtesy of Dr. W. E. Nickola.*)

resonant frequencies for the circular membrane in vacuum shows a good agreement[19] and therefore demonstrates the reliability of linear membrane theory for circular specimens. In air, the amplitude versus frequency resonant peaks in the experiments were

[19] Nickola, W. E., *Exp. Mech.* **6** (12) 513 (1966).

very sharp and moiré fringes were detectable over a frequency range not exceeding 8–10 Hz at each nodal frequency. It was also experimentally proved that each resonant frequency was independent of the vibration amplitude of the membrane.

However, the moiré patterns created by this technique and shown in Fig. 7.9 are in reality slope contour patterns of the flexed membrane at resonant frequencies, frozen by the aid of the intermittent light emitted by the two stroboscopic light sources when the membranes were at maximum deflection. These are not nodal patterns as the so-called Chladni's figures which depict the nodal distribution of a vibrating membrane or plate at resonance.

7.2. The moiré methods applied to transient strain distributions

Besides the well-known techniques of electrical strain gauges and condensers extensively utilized for the determination of transient strains on the surface of metallic plates and bars, optical methods have also been used in studying transient problems. These optical methods include the optical interference, the photogrid and the photoelastic techniques. While the first method is extremely sensitive, it necessitates costly and very delicate instruments (interferometers) and laborious and skilful techniques for adjusting the experimental set-up, the second technique is a rather coarse technique yielding qualitative results. The photoelastic method is limited by the necessity of using transparent birefringent materials by the fact that the photoelastic effect is an integral effect along the path of the light crossing the specimen and by the necessity of auxiliary information from other methods for the separation of the principal stresses.

The moiré method may be used to complement photoelasticity and yield complete information about a transient stress field. As such it has been applied by Durelli et al.[20-22] for the study of

[20] Riley, W. E., and Durelli, A. J., *Trans. ASME, J. Appl. Mech.* **29**, 23 (1962).

[21] Riley, W. F., and Durelli, A. J., *Int. J. Mech. Sci.* **2** (4) 213 (1960).

[22] Durelli, A. J., Riley, W. F., and Carey, J. J., *Proc. Int. Symp. Photoelasticity*, M. M. Frocht (Ed.), Pergamon Press, Oxford, 1961, p. 251.

transient stress distribution of plates made of a low modulus urethane rubber. The problems studied were related to the determination of (a) the magnitudes and directions of principal stresses of a generic point of a circular disc subjected to a time dependent loading at a point on one edge of the disc, and (b) the stress distributions on the boundary of an elliptical or a square hole in large plates during the passage of a stress wave of long duration. The problems dealt with the classical photoelastic methods for the determination of the difference of principal stresses complemented by the linear differential moiré method for evaluating the normal components of strains and stresses at each point of the field.

Since the time-dependent load was applied to the axis of symmetry of the specimen, the two specimen gratings were attached to the two halfs of the plate and were right angles. The specimen gratings were reproduced by contact printing on one lateral surface of the plate from a master grating having a pitch 0.001 (4λ) in. A polyvinylalcohol enamel known commercially as Gaco was used as photosensitive material. The fixing process of the photosensitive material was not completed since the soft plate could not undergo the high temperatures of the baking process of the material. Thus the reproduced specimen grating remained in blue instead of black colour, but this difference did not harm the quality of moiré patterns. The reference grating of a frequency of 1000 lines per inch was reproduced on a flat glass plate.

The loading of the specimen was accomplished by dropping a weight on a hard plastic striker which was permanently positioned on the specimen. The reference grating was placed in close proximity to the specimen grating parallelly to the surface of the unloaded specimen. The small air gap was filled with mineral oil so that the oil-film kept the reference grating in contact with the plate.

Photographic recording was achieved by a microflash as a light source and a still camera. The flash was triggered by a signal from a barium–titanate pressure gauge embedded in the striker. A special timer unit was used to delay the flash for a predetermined delay

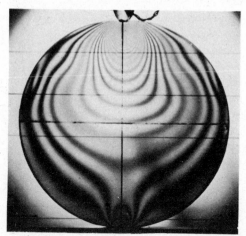

PHOTOELASTIC LIGHT FIELD FRINGE PATTERN

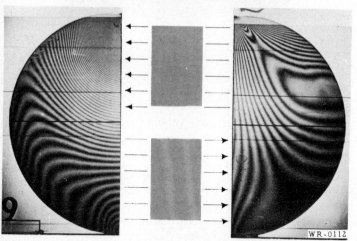

MOIRE FRINGE PATTERNS

FIG. 7.10. Isochromatic light field fringe pattern of a disc subjected to diametric impact loading and moiré patterns of the same disc along the two principal directions of the specimen. (*Courtesy of Prof. W. F. Riley.*)

interval, the duration of which was measured by a counter. Figure
7.10 shows the isochromatic fringe pattern of the disc in light field
as well as the two moiré patterns along the two principal directions
of the specimen.

The interpretation of the photoelastic and moiré patterns requires
a number of patterns to be taken over relatively short intervals in
order to determine the strain rates. This necessitates a reproducible
loading. For the correlation of the data obtained from moiré
patterns, which yield components of strain, with the photoelastic
data referred to components of stresses, an iterative process was
introduced where at the beginning an arbitrary value of the
modulus of elasticity was selected and all the desired quantities
were computed for a number of times during the interval of
interest. The rates of change of the principal strains were then
computed and used to obtain a better estimate of the modulus at
each instant. By repeating this iterative procedure several times the
correct value for the dynamic modulus of elasticity may be
evaluated.

The results obtained from this study indicate that the moiré
method can be effectively used to complement the classical photoelastic method in transient problems of stress and strain distribution.

Finally, it should be mentioned that the linear differential moiré
method is self-sufficient when applied to dynamic problems. If a
third grating oriented to 45° relatively to the two initial gratings is
printed on the specimen it provides a third moiré pattern which,
together with the two initial right-angle patterns, yields complete
data for the evaluation of the problem. In this case the loading
must be reproducible because the two or three moiré patterns must
be obtained in independent loadings.

7.3. The linear differential moiré method applied to shock
wave propagation

As a third example of the application of moiré methods to
problems of dynamic strain analysis, a study of the propagation of

shock waves in transparent spheres made of Perspex was chosen. Applications of the differential moiré method to shock wave propagation problems were made by Theocaris *et al.*[23] The problem presents a radial geometric and loading symmetry and therefore the linear differential moiré method, as developed in § 2.4, gave satisfactory results.

The shock wave was produced by detonating an explosive charge in the central cavity of Perspex spheres. The explosive charges consisted of Pentolite spheres made from the same mould for all charges.

The charges were detonated with M36 detonators set off by a 5 kV pulse from a 4 μF capacitor, which was fired by a thyratron circuit. The size and type of charge was selected such as to produce moderate pressures during explosion. At these pressures the pressure versus relative volume curve for Perspex presented an inflection point due to relaxation phenomena engendered during the explosion in Perspex, which resulted in a wave structure consisting of an elastic wave or precursor followed by a plastic shock wave with a propagation velocity much lower than that of the precursor.[24]

The diameters of the Perspex spheres were chosen such as to ensure that eventual geometric discontinuities did not affect the assumption of an explosion at the interior of a homogeneous medium. Five spheres were tested, the diameters of which varied between 7·373 and 7·341 in. The spheres were fabricated from two hemispheres. A line grating of a frequency of 200 lines per inch, reproduced on a stripping film, was cemented on to the polished flat surface of one hemisphere of each sphere. After the setting of the cement the film was stripped from its base and the two parts of each sphere cemented together with the same cement. After the polymerization process of the cement the specimens were ready for

[23] Theocaris, P. S., Marketos, E., and Gillich, W., *Int. J. Mech. Sci.* **8**, 739 (1966), and Theocaris, P. S., Davids, N., Gillich, W., and Calvit, H. H., *Exp. Mech.* 7 (5) 202 (1967).

[24] Duvall, G. E., Stress waves in anelastic solids, *Proc. IUTAM*, H. Kolsky and W. Prager (Eds.), Springer-Verlag, Berlin, 1964, p. 20. See also Jacquesson, J., *ibid.*, p. 33.

testing. The bond proved to be satisfactory as all spheres withstood the blast wave and ruptured in places different than the cemented sections.

For the measurement of the characteristic quantities of the multiple shock wave propagated in the Perspex spheres the linear differential moiré method was used. The image of the specimen grating at unit magnification was received at the ground-glass screen of a still camera, which interfered with a reference grating of the same frequency as the specimen grating, placed on to the ground-glass screen of the camera. Slight modification of the distance between the lens of the camera and the reference grating

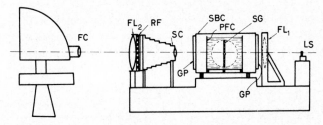

FIG. 7.11. Apparatus used for recording moiré patterns in exploding Perspex spheres.

formed an initial moiré pattern with parallel and equidistant fringes. The initial moiré pattern introduced a fictitious initial compression in the specimen.

The apparatus, shown in Fig. 7.11, consisted of an exploding wire light source LS, collimated by a Fresnel lens FL_1. A steel blast container SBC having two glass ports and filled with silicone fluid housed the spheres. The refractive index of the silicone fluid matched the refractive index of the Perspex sphere before explosion. A still camera SC with the reference grating juxtaposed on its ground-glass screen formed the moiré pattern of the meridian section of the sphere. A framing camera FC recorded the explosion phenomenon. A Beckman and Whitley No. 192 continuous writing framing camera was used. Different registration speeds were used in order to study the various stages of explosion. In this discussion only the initial stage is examined, where the exploding gas comes into contact with the solid and generates a shock wave, up to the point where the shock wave reaches the external surface of the sphere.

Figure 7.12 shows two characteristic phases of the propagation of the shock wave in the sphere. While Fig. 7.12a shows the initial moiré pattern at the instant when the explosion starts, Fig. 7.12b

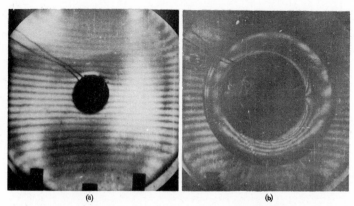

(a)   (b)

Fig. 7.12. Moiré patterns of two characteristic phases of the propagation of a shock wave in a Perspex sphere. (a) Initial moiré pattern at the instant of explosion. (b) Moiré pattern of the shocked Perspex sphere when the wave front had advanced deep in the interior of the sphere.

presents a typical case of the shocked Perspex sphere when the wave front had advanced deep in the interior of the sphere. The refractive index of the material in the shocked zone changed continuously because of the density variation in Perspex. Therefore there was a continuous deviation of the beams of light passing through the shocked sphere. This resulted in the creation of dark zones which appear in the moiré pattern. The number of dark zones corresponded to the discontinuities of the shock wave. Without defining their exact position these zones indicated the decomposition of the shock wave into two parts, i.e. a precursor elastic wave and a following plastic shock wave. The dark zones were broad at the beginning of the explosion and their width steadily diminished as the explosion propagated outwards due to the reduction of sphericity of the shocked part of the sphere.

Behind the wake of the precursor the material returned to a regime comparable to its undeformed state. A moiré pattern

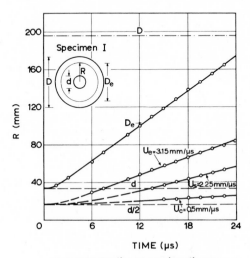

FIG. 7.13. Radius versus time diagram (R, t) corresponding to the propagation of the precursor and plastic shock wave fronts, as well as to the expansion of the internal cavity of an exploding Perspex sphere.

appeared behind the shock wake which yielded the u-displacement component along a direction normal to the rulings of the grating.

From the series of moiré photographs taken by the framing camera, diagrams of variation of the precursor wave and the plastic shock wave fronts were plotted versus time, as well as the expansion of the internal cavity. Figure 7.13 shows the (R, t) diagrams of these three quantities. It can be deduced from this figure that the velocities of propagation of the shock waves were constant and they diminished from the precursor to the shock wave. Similarly, the expansion of the internal cavity of the sphere was linear with time.

A main feature of the moiré method is that it yields a means for a direct and continuous observation of the wakes of the propagating shock waves during the evolution of the explosion phenomenon without necessitating the use of various observation stations for the measurement of the time of passage of the shock wave, as is customary with all other experimental methods for the detection of the propagation of shock waves.

The velocity of the precursor wave U_e, as it was measured by the moiré pattern ($U_e = 0.310$ cm sec^{-1}), compared well with the value derived from the mechanical properties of Perspex at its glassy state (relaxation modulus $E_g = 42.20 \times 10^9$ dyn/cm^2, Poisson's ratio $v_g = 0.420$ and the density of the undisturbed material $\rho_0 = 1.14 \times 10^{-3}$ g.cm^{-4} sec^2) and the relation expressing the velocity propagation of elastic disturbances

$$U_e = \frac{(1-v)}{(1+v)(1-2v)} \frac{E}{\rho_0}.$$

The velocity of the precursor wave was found to be $U_e = 0.310$ cm sec^{-1}. This agreement indicates that the material at the wake of the precursor is still at its glassy state and no significant changes to its viscoelastic behaviour have yet taken place.

The plastic shock wave velocity, which was measured to $U_s = 0.225$ cm/μsec, corresponds to a state of the material derived from the following relation:

$$U_s^2 = \frac{K}{\rho},$$

where K is the bulk modulus of the material and ρ the density at the shocked state of the material. Accepting an average value for the density at the plastic shock of $\rho = 1.16 \times 10^{-3}$ g.cm^{-4} sec^2 it is found that $K = 57.50 \times 10^9$ dyn/cm^2. This value for bulk modulus corresponds to $E = 23.40 \times 10^9$ dyn/cm^2 and a value for Poisson's ratio $v = 0.432$. These values show that the material, when the plastic shock wave passes, has already undergone some change in viscoelastic behaviour.

The radial and tangential components of strains were evaluated behind the precursor and the shock wave by measuring the radial displacement u and its variation along a radius of the sphere at different time intervals. It was assumed during this evaluation that the sphere was deformed in a radial direction and that the displacement only depends on radius r and time t. This assumption is justified by the moiré patterns which show a complete symmetry of the displacement field. Since the moiré pattern expresses the displacement in a direction normal to the rulings of the gratings, no fringes appeared in a direction normal to initial fringes (Fig. 7.12b). Another assumption made, which is again justified by the experimental evidence, is that the displacement of a generic point of the sphere depends only on the influence of the shock, which has already passed the point in question and there is no, or insignificant, influence of the shock wave behind the point.

A third assumption made is related to the influence of difference of refractive index. Although great care was taken to match the refractive index of the immersion oil with that of Perspex, there was a change in density of the shocked part of the sphere and therefore of its refractive index. The parallel beam of light passing through the shocked part of the sphere with higher density and different refractive index deviates from parallelism and produces the dark regions at the vicinity of the shock waves. This deviation increases as the diameter of the shocked sphere decreases due to the increasing sphericity.

Another effect due to the variation of the refractive index is the abrupt change in interfringe spacing of the moiré pattern behind the wake of the precursor. In order to evaluate the correction factor for the variation of the refractive index, it was assumed that there was no tangential displacement at the wake of precursor. Then the moiré pattern in a direction parallel to the lines of the specimen grating (i.e. on a horizontal axis passing through the centre of the spheres in Figs. 7.12a, b) must be continuous before and after the wake of the precursor. The difference between the moiré fringes before and after the precursor wake is due to the influence of the variation of refractive index. The corrections made were based on this assumption and the values of the displacement u were referred to an initial moiré pattern eliminating the u_t along the wake of the precursor.

This correction derived from the moiré pattern is more than satisfactory for the limited region from which the experimental

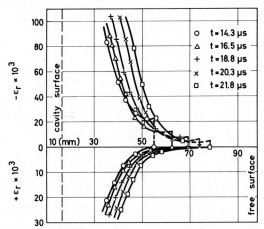

FIG. 7.14. Radial ε_r and tangential ε_t strain distributions along a radius of an exploding Perspex sphere at different time instants.

data were derived by the moiré technique. Indeed, the region of measurements used for the results was restricted to a quarter of the radius around the mid-radius of the sphere. It was observed that the deviation due to sphericity in this narrow ring from which the data was taken was insignificant. This can be checked from Fig. 7.12a of the initial moiré pattern ($t = 0$), where the deviation of the moiré fringes in this region due to sphericity is negligible in comparison with the order of magnitude of the displacements measured by the moiré pattern. The overall experimental error was estimated to be of the order of 10–15 per cent, which is very satisfactory for this type of test.

By measuring the component of the radial displacement, as well as its variation along the radius of the sphere, at each time interval, from the corresponding moiré pattern the ε_r and ε_t components of strain can be determined by applying the relationships

$$\varepsilon_r = \frac{\partial u}{\partial r} \quad \text{and} \quad \varepsilon_t = \frac{u}{r}.$$

These relations holding for small strains were justified from the moiré patterns, the density of the fringes of which indicated small

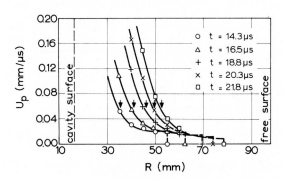

FIG. 7.15. Particle velocity U_p distributions along a radius of an exploding Perspex sphere at different time instants.

displacements and displacement variations along a radius of the sphere.

Figure 7.14 shows the distribution of radial and tangential strains ε_r and ε_t as functions of the radius at different instants. While the increase of the tangential strain ε_t, which is extensive, starts from zero at the wake of the precursor, the radial compressive strain ε_r presents a jump in ordinate at the precursor wake. Both components of strain increase slowly and smoothly up to the shock wave, where a steady and steeper increase of both strains appears.

The radial displacement differentiated with respect to time yields the particle velocity U_p. The distribution of the particle velocity along the radius of the sphere for different time instants is given in Fig. 7.15.

The determination of shock wave velocity, particle velocity and internal cavity velocity, as well as the components of strain yield sufficient information for the complete determination of the state of each point behind the shock wave and completely solves the problem of pressure distribution and density variation at the shocked area of the body.

In these three illustrative examples, where problems of strain analysis were treated by applying different moiré techniques, it was shown that the moiré method is very effective in solving problems presenting geometric and loading symmetries. Since in most cases of practical applications such symmetric cases generally exist, it is admissible to accept the general validity of the moiré method to any problem of strain analysis. In the very special cases where the body does not present any form of geometric symmetry and the external loads are unsymmetrically applied, it is possible, as it has been previously mentioned, to use an extra specimen grating oriented at some angle to the two basic orthogonally oriented gratings which will yield additional information sufficient to ascertain the accuracy of the results derived from moiré patterns.

CHAPTER 8

The Moiré Methods applied to Curved Surfaces

THE theory of moiré fringes has extensively been applied to the study of the topography of flat plates and beams. These methods, as they were described in previous chapters, are limited to cases of flat surfaces. The reflected image of a slightly curved grating, because of the deformation of the flat plate, slightly deviates from the original flat surface. Thus it can be brought into focus on the photographic plate of the camera whilst simple linear relations connect the quantities measured by the moiré fringes.

The simplicity in recording either deflection or partial slope contours in flat plates by the moiré fringes is not maintained in the cases of curved surfaces. This is because the reflected image of a grating on a curved surface does not lie in one plane and focusing is impossible. Moreover, in the case of flexed flat plates the initial curvature is approximately equal to zero and hence of the same order as the small changes of curvature generated by flexure. In the case of shells the initial curvature may be of much higher order than the change of curvature developed by flexure. Finally, for each case of shell having a distinct curvature a conjugated grating of special geometry is necessary in order to avoid complexities in the determination of the quantities to be measured. For an arbitrary shape of the conjugate grating, deflections. slopes and curvatures at each point of the shell are interrelated, and complicated computations may be required for the separation of these quantities. Then the correct shape of the conjugate grating is directly related to the dimensions and curvature of the specimen as well as to the geometry of the apparatus used. Therefore for each case

312

studied a special arrangement of the apparatus and a special shape for the reference grating is needed. This complicates considerably the application of the method to curved surfaces.

De Josselin de Jong[1] and Evensen[2] have developed experimental methods which, besides those described in this chapter, deal with the applications of moiré fringes to curved surfaces.

De Josselin de Jong used a flat reference grating which was projected through a transparent twin specimen on to a second grating placed behind the specimen. Moiré fringes yielded the partial slope contours when an opposed load was applied to the two parts of the specimen. Unidirectional loading of the twin specimen yielded the so-called "isoflexes" and "isotorses". The method is rather complicated and is applicable only to curved surfaces of small curvature.

Evensen, by observing the reflected image of a grid-shaped light source from the polished surface of a cylinder, investigated the phenomenon of elastic buckling. The image is sensitive to changes of curvatures but yields only qualitative results.

Besides these methods surveying curved surfaces, there are different standard optical techniques which are appropriate for the measurement of curvature changes in small flat or curved surfaces. The interferometer by Gates[3] is appropriate for the measurement of microscopic deviations in small spheres. A light beam is directed normally on all points of a curved concave reference surface and is reflected from the convex sphere under study. The specimen is brought into central coincidence with the concave lens and the interference fringes yield departures from sphericity.

Einsporn's[4] and Martinelli's[5] spherometers measure the curvature of convex surfaces by focusing a reflected collimated light

[1] De Josselin de Jong, G., *Proc. Symp. Shell Res.*, North Holland Publ. Co., Delft, 1961, p. 302.

[2] Evensen, D. A., *Exp. Mech.* **4** (1) 110 (1964).

[3] Gates, J. M., Optics in metrology, *Coll. Int. Comm. Opt.* 1958, Pergamon Press, Oxford, 1960, p. 201.

[4] Einsporn, E., *Z. Instrum.* **57**, 267 (1937).

[5] Mindlin, R., and Salvadori, M., Analogies, *Handbook of Experimental Stress Analysis*, M. Hetényi (Ed.), Wiley, New York, 1950, p. 787.

beam on the surface of the specimen through a lens of known focal length and distance from the specimen. Martinelli's spherometer equipped with a Ronchi grating placed in a position near the focus of the lens enables easy adjustment of the true focus by counting the number of fringes visible in the field of view of the instrument. The Martinelli–Ronchi spherometer has been introduced by Duncan and Sabin.[6]

The above-mentioned spherometers are single point mechanical devices for measuring the z-coordinate or the curvature of a curved surface from a reference surface. Therefore, the procedure of mapping the topography of surface becomes a point-by-point process which is time-consuming and rather inaccurate.

In this chapter two moiré methods will be introduced for the study of curved surfaces. These methods are extensions of the previously described methods of the oblique shadow for the evaluation of the deflection contours and of the reflected image method for recording the partial slope contours in flexed plates (see §§ 6.2 and 6.6). Both methods, in their simplest form, are applicable to axisymmetric developable or ruled surfaces. They can be used to non-axisymmetric surfaces and generally to any type of surface, but in this case they lose their simplicity.

A great part of surfaces of practical interest are basically spheres, cylinders, cones and ruled surfaces such as hyperboloids. All these surfaces can be studied by the moiré methods developed in this chapter.

For the case of spheres it is necessary to use, instead of plane reference gratings, gratings wrapped on a cylindrical surface of a slightly greater radius than that of the sphere and with their rulings normal to the axis of the cylinder.

The first method, called the oblique-shadow method (OBS), employs diffuse reflection and therefore requires a matte surface finish on the specimen. Thus, the method avoids the requisite of polishing surface, which is a great advantage. A reference grating RG_1 is placed in contact with either a generator or a straight line of the surface and is illuminated by a parallel light beam.

6 Duncan, J. P., and Sabin, P. G., *Exp. Mech.* **3**, 285 (1963).

nterference of the rulings of RG_1 and their shadows on the matte reflecting surface yields the deflection contours of the shells.

The second method, called the reflected image method (RIM), requires the specimen surface to be specularly polished. The reference grating RG_1 is placed at a distance from the surface and the reflected image of RG_1 on the polished surface of the shell is brought to interfere with a second grating RG_2, similar to RG_1. The method yields partial slope contour moiré patterns of the deformed shell.

Both methods are capable of measuring either the macroscopic shape of a curved surface by defining their radii of curvature, as well as their variations, when the surface remains in its natural state, or the microscopic shape when it is deformed by externally applied loads.

In both methods and in the case where the microscopic variations of the curvature of the shell due to an external load are studied, the specimen is rotated on a turntable in order to eliminate the influence of the initial curvature of the shell. In this way narrow strips of the curved surface are successively brought to the reference position and photographed by a periphery camera. The lateral surface of the shell is, in this way, replaced by the lateral surface of a prism inscribed to the shell which has an infinite number of lateral faces.

However, the same set-up may be used for the measurement of the initial curvature in axisymmetric shells and the variations of curvature in non-axisymmetric shells due to their shape, as the set-up used in the case of measurement of microscopic variations of curvature due to loading. The only difference between the two arrangements is that, in the case of measurement of initial curvatures, the apparatus is adjusted so that the photograph of the lateral surface of the shell taken by the periphery camera does not show moiré fringes. From the characteristics of the set-up at zero moiré fringe position and the pitch of the reference grating the initial curvature of the shell can be determined.

Then both methods are capable of surveying curved surfaces of any size and dimension in their natural or deformed state without necessitating a special arrangement of the apparatus and a special shape for the reference grating conjugated with the shape and curvatures of the specimen shell. These are certain advantages of the methods which make them easy to handle and quasi-universal for the main types of shells used in practice. Both methods were introduced by Theocaris.[7]

Ligtenberg's photoreflective moiré method can be extended to the study of deformed curved surfaces. This idea was introduced by Duncan.[8] He employed the idea set up by Theocaris[9] to use a periphery camera for depicting the partial slope contours in cylindrical surfaces by unwrapping the cylindrical surface into a plane. In order to reveal small departures from the initial cylindrical shape due to loading it is necessary to develop a uniformly pitched curved ruled screen conjugated to the characteristics of the cylindrical surface under study so that moiré fringes formed by superposition of the original and the loaded patterns show a constant slope interval.

The problem essentially lies on computing the polar coordinates of elementary strips of the conjugate screen related to a particular cylindrical surface. These strips are parallel to the generators of the cylinder and the whole screen is concentric with the axis of rotation of the cylindrical shell. The screen is ruled with equispaced black and white lines of equal pitch parallel to the generators of the cylinder. The screen is designed to be viewed from a certain point carrying a hole for photography, through reflection from the cylindrical surface.

The tangent plane at each generator reflects a particular line of the screen when the shell is undeformed. The shape of the screen is such that, while it is ruled with equispaced lines, the tangent plane at each point of the screen displaces the image by a fixed

[7] Theocaris, P. S., *Explle Spannunganalyse*, VDI-Berichte, No. 102, VDI Verlag, Dusseldorf, Germany, 1966, p. 44. See also, Theocaris, P. S., *Exp. Mech.* 7 (7) 289 (1967).

[8] Duncan, J. P., *The Optical Survey of Curved Surfaces*, Dept. of Mech. Engrg. Rept., Univ. Brit. Columbia, Canada, 1966, ch. 9.

[9] Theocaris, P. S., *Exp. Mech.* 7 (7) 289 (1967).

multiple of the screen pitch for an arbitrary fixed local angular displacement of the tangent plane. Then, superposition of the original and the loaded patterns of the screen lines give a moiré pattern, the fringes of which represent constant increments in angular displacement of tangent planes at generators.

The shape of the ruled screen for a particular cylindrical surface (8 in. in radius) was developed in a digital computer. The method seems complicated since it necessitates the definition of a conjugate screen for each curved surface under study.

A very promising method for the study of slope contours in deformed bodies of any shape is the technique based on holography. This technique is still in its infantile stages and for this reason only the basic principles of this method will be discussed at the end of this chapter.

8.1. The oblique shadow moiré method for the study of curved surfaces

A physical surface presents a microscopic curvature and a macroscopic curvature associated either with roughness or with deformation of the surface due to external loading. The moiré technique developed in this section deals with the measurement either of the macroscopic curvature or of variation of partial slopes and deflections in curved surfaces. This technique is therefore capable of measuring both the undeformed shape of the shell as well as small departures from the initial shape due to loading.

If the macroscopic shape of a physical surface is concerned and any kind of roughness and local perturbation of the physical surface about a mean theoretical shape is disregarded, a curved surface may be defined in terms of Cartesian coordinates. The xy plane is considered tangent to a generic point O of the surface, while the z-axis coincides with the normal to O (Fig. 8.1). In this case the surface can be studied in terms of meridional sections referred to the tangent xy plane. Deflection w along the z-axis for each point is due to the macroscopic deformation of the surface originated by loading. Then deflection w may be mapped as a family of deflection contours projected on the reference plane xy, each contour line having a difference in elevation h with the two neighbouring contour lines.

On the other hand, the slope of the surface along an arbitrary direction n forming an angle ω with the x-axis is given by

$$\frac{dw}{dn} = \frac{\partial w}{\partial x} \cos \omega + \frac{\partial w}{\partial y} \sin \omega. \qquad (8.1)$$

If point P along the direction n is sufficiently close to origin O, so that the length of the arc δs is approximately equal to the

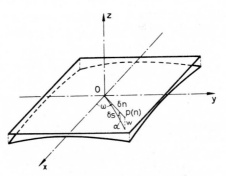

Fig. 8.1. Schematic diagram of a curved surface and its tangential plane at zenith O of the surface.

distance δn, then the derivatives $\partial w/\partial x$, $\partial w/\partial y$ and dw/dn express the angles α formed between the axes of the reference plane x, y and n and the corresponding curves of the surface. In this case the corresponding curvature k may be approximated,

$$k = \frac{d\alpha}{ds} \approx \frac{d^2w}{dn^2}. \qquad (8.2)$$

Distance δs must be taken very small for highly curved surfaces and can be relatively large for nearly flat surfaces in order to satisfy the above approximation.

Since surfaces of practical importance are those of developable shells (cylinders and cones) the shell specimen may be considered as a developable surface and it can be divided into narrow strips having a length equal to that of the specimen and an infinitesimal width b. The width b depends on the relative magnitude of the radius of curvature R and the change of deflection Δw due to the initial curvature of the shell and its deformation during loading. If the gap between the curve corresponding to the physical surface of the shell and its tangent at a generic point O, which may be taken as the zenith of the curved surface, is small, the curved strip may be approximated to a plane surface. By properly selecting width b, the validity of the approximation of a flat strip can be accepted and the curved surface may be regarded as formed by a succession of flat strips of different width depending on the particular curvature of the part of the surface belonging to each strip. Then the physical surface is replaced by a lateral prismatic surface equivalent to the curved surface. If each face of the prismatic surface is brought successively into the same position as its predecessor the initial curvature of the surface can be ignored and the variation of curvature and slope of the surface due to loading can be studied.

The idea of bringing successively narrow strips of the curved surface to a reference position and thus developing the surface under investigation into a plane surface has been realized by using a "periphery camera".[10] The camera is shown in Fig. 8.2 and is capable of photographing external or internal surfaces of cylindrical or conical shells whose diameters and ratios of diameters to height may widely differ, as well as shells with appreciably different forms from those of circular cylinders. The camera has been designed as two separate units: a rotating table carrying the specimen and bringing successively narrow strips of the cylindrical or conical surface to the centre line of the camera and a camera embodying a mechanism to move the film behind a stationary slit in step with the moving image of the rotating specimen. The control of the speed of rotation of the turntable and translation of

[10] Trade name of a camera sold by Research Engineers Ltd., London, England.

the film is achieved by driving both camera and turntable mechanisms by synchronous motors. Since immediately in front of the film the narrow slit S is present, the film sees only a narrow central

FIG. 8.2. Periphery camera and its accessories.

strip of the specimen. As the specimen rotates its image moves in a direction shown by the arrow in Fig. 8.9a (p. 315).

In order to create a sharp picture, the image of the rotating specimen must keep in step with the movement of the film. The magnification which satisfied this requirement is given by the simple relation

$$\frac{l_4}{l_5} = \frac{2\pi r}{t},\tag{8.3}$$

where r is the radius of the image of the cylinder at RG_2, t the traverse of the film and l_4 and l_5 the distances of the principal point of lens L_2 from RG_2 and the film.

If the difference between maximum R and minimum r radii is δr and the camera is set up for a pure cylinder of radius r, the image of the smaller cylinder moves at the speed of the film. The image of the larger cylinder moves faster, thus increasing the local rate of developing of the surface. Hence the image at a point on the negative corresponding to a radius R represents a larger per unit time developable area. This phenomenon results in a blurring of the image. In order to avoid blurring, the images of parts of the specimen with different radii must keep sensibly in step over the width of the slit. In the time t taken by the image of the smaller cylinder to traverse slit S the image of the larger cylinder will move a distance $s\dfrac{R}{r}$, therefore difference $(s\dfrac{R}{r} - s) = s\dfrac{\delta r}{r}$ must be kept smaller than a limit value. The slit of the camera is adjustable and in the case where there is large difference between R and r this slit may be adjusted very narrow to suit the circumstances.

For the case of circular cylinders loaded elastically or plastically the difference between R and r is insignificant, but in cases of non-axisymmetric shells the slit may be adjusted to conform with the requirements of sharpness of the photograph. For axisymmetric shells the size of the slit must vary according to the requirements of planeity of the image of the shell which, in turn, depends on the initial curvature of the shell.

It is clear from the above discussion that the same technique can be applied to any type of axisymmetric developable or ruled surface independently of the geometric characteristics of the particular surface, provided that a judicious choice is made for the limiting value of width b of the elementary strip, as well as for the width of slit S.

The apparatus used for the application of the method is shown in Fig. 8.3. A parallel white-light beam produced by light source LS illuminated obliquely, at an angle of incidence i, the matte surface of the shell. A reference grating RG was placed between LS and specimen Sp. By viewing the surface in a normal direction

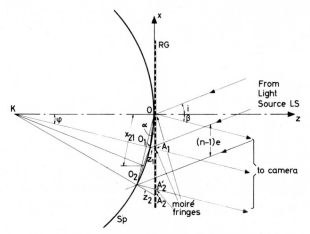

FIG. 8.3. Apparatus used in oblique shadow moiré method.

through RG a moiré pattern appeared, formed from the interference of the grating and its shadow. The moiré pattern depicted the topography of the shell surface. The interval in height Δz between two successive moiré fringes may be shown from simple trigonometry to be

$$\Delta z = \frac{p}{\tan i}, \qquad (8.4)$$

where p is the pitch of RG.[11]

The departure Δz from RG between any pair of successive

[11] Theocaris, P. S., *Proc. Intern. Symp. Shell Structures, Warsaw*, 1963, North Holland Publ. Co., Amsterdam, 1965, p. 887.

fringes contains, besides a term due to the deflection of the surface Δw, another term due to the curvature of the shell Δc, $[\Delta z = \Delta w + \Delta c]$. By reducing the area of the shell surface to a narrow strip of width b the quantity Δc was reduced to an insignificant fraction of the total deflection Δz. Thus $\Delta z \approx \Delta w$ and the moiré pattern shows the deflection-contour map of the shell. Even in the very seldom case where width b cannot be reduced sufficiently so that $\Delta c \ll \Delta w$, the unloaded pattern of the surface of a developable axisymmetric shell will contain parallel and equidistant fringes because of the constant curvature at sections parallel to the base of the shell. This pattern, when subtracted from the loaded pattern, will yield the deflection contour pattern due to loading.

Let us consider the case of a cylindrical shell of a diameter $D = 3 \cdot 45$ in. (84 mm). For a slit of the periphery camera equal to $0 \cdot 0075$ in. ($0 \cdot 19$ mm), which is satisfactory to produce sharp and well-contrasted fringes with a simple system of illumination, width b equals $0 \cdot 02$ in. ($0 \cdot 5$ mm). This width gives

$$\Delta c = 0 \cdot 000029 \text{ in. } (0 \cdot 0007 \text{ mm}).$$

If the reference grating has a frequency of 200 lines per inch and an angle of incidence $i = 20°$, the departure from the reference plane RG between two successive fringes is given by relation (8.4):

$$\Delta z = 0 \cdot 014 \text{ in. } (0 \cdot 35 \text{ mm}).$$

Then Δc is approximately one five-hundredth of Δz and the error in neglecting the curvature of the strip of the cylinder of a width $b = 0 \cdot 02$ in. is less than $0 \cdot 002$.

In the case of conical shells the turntable of the periphery camera must be tilted to an angle equal to half angle of the cone, so that the generator of the cone facing RG_1 remains always parallel to the plane of RG_1 during rotation. The shape of the lateral surface of the cone will then be distorted because each point of a generator of the lateral surface will move at different speeds between zero at the apex and the maximum speed for the points at the cone's base.

But, since the speed variation with height is linear, a correction of the moiré pattern can be readily obtained.

Similarly, in the case of a ruled surface (e.g. a hyperboloid) it must be simultaneously rotated by two constant rotations. The developed surface of the shell corresponds to a distorted image of the physical surface of the shell. However, by knowing the instantaneous angles of rotations of the shell during photographing, it is possible to define the correspondence of the relative points between the physical surface and its distorted image made by the photographic process.

For the evaluation of the geometric initial curvature of a physical surface, the width of slit S of the periphery camera must be increased progressively up to the point where a moiré pattern of the undeformed cylindrical or conical shell is apparent at the photograph. This moiré pattern contains equidistant moiré fringes parallel to a generator of the shell. For the case of cylinders these fringes will also be of equal width, while for cones these fringes will be of linearly variable width from the base to the apex of the cone. These fringes are derived from the initial curvature of the shell and they therefore allow the evaluation of the radius of curvature of the shell.

Considering the surface of the cylindrical shell to be smooth at the vicinity of zenith O where the truly cylindrical surface is tangent to the reference plane RG, the distances $OO_1 = x_1$, $OO_2 = x_2$ correspond to the mid-lines of successive moiré fringes formed by the interference of the RG and its shadow on the matte cylindrical surface. The departures $O_1A_1 = z_1$, $O_2A_2 = z_2$ of the cylindrical surface at points O_1, O_2, etc., from the reference plane may be taken equal to the corresponding distances O_1A_1', O_2A_2' along the direction of viewing of the camera which subtends an angle β to the z-axis because the analysis is restricted to a narrow strip of the cylinder extending on both sides of the generator passing through the zenith O.

If the normal to the side OO_1 in the triangle KOO_1 (Fig. 8.4) is traced, this normal cuts the Ox axis at point A_0. OL is normal on KA, and from the orthogonal triangle KOA_0, where

$KL \approx [R - (z/2)]$, $LA_0 \approx z/2$ and $OL = x/2$ it can be deduced that

$$x_1^2 = z_1(2R - z_1) \approx 2Rz_1.$$

Similarly $x_2^2 = z_2(2R - z_2) \approx 2Rz_2$

and therefore $x_2^2 - x_1^2 = 2R(z_2 - z_1) = 2R\Delta z.$ (8.5)

This relation holds for small departures of curvature ($\Delta c \equiv \Delta z$)

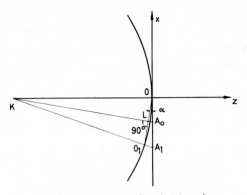

FIG. 8.4. Geometry of a local region
of the cylindrical shell at the neighbourhood of zenith O.

as compared to the radius R of the shell. Then, from relation
(8.5), it is valid that

$$R = \frac{(x_2 - x_1)}{\Delta z} \cdot \frac{(x_2 + x_1)}{2}.$$ (8.5a)

If the distance from the middle point between two successive
moiré fringes (light fringe) and the zenith O is called $x_{21} = \dfrac{x_2 + x_1}{2}$
and the interfringe spacing between successive moiré fringes
$\Delta x_{21} = (x_2 - x_1)$ then relation (8.5a) becomes

$$R = x_{21}\frac{\Delta x_{21}}{\Delta z}.$$ (8.5b)

Applying this relationship to the two first successive dark fringes passing through the zenith O and point O_1 it may be readily found that

$$R \approx \frac{x_{21}}{\tan \alpha},\tag{8.6}$$

where α is the angle subtended between the tangent plane RG and the chord OO_1 of the cylindrical surface between two successive moiré fringes.

The field within which the validity of approximations in expressions (8.5b) and (8.6) is satisfactory depends upon the relative magnitude of R and Δz. With R large and Δz very small the gaps between successive OA_1 and OO_1 are small. The curved surface may be approximated to a plane surface enclosing with the generator of the cylinder at the zenith (y-axis) a wedge and therefore the zone of interest in the cylinder must be confined to a narrow excursion from the zenith O.

Since, generally, it is valid that

$$\tan \alpha = \frac{z}{f} \quad \text{and} \quad z = \frac{p}{\tan i},$$

then

$$\tan \alpha = \frac{p}{f \tan i},\tag{8.7}$$

where f is the interfringe spacing between two successive moiré fringes. Introducing eqn. (8.7) into eqn. (8.6) the radius of the shell is

$$R \approx x_{21} \frac{f \tan i}{p}.\tag{8.8}$$

This relation yields the initial value of the shell radius from quantities measured on the moiré pattern. However, the accurate measurement of the quantities in relation (8.8) is rather difficult and the values of R derived by this method present some error.

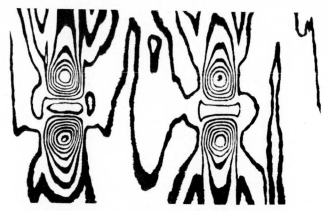

FIG. 8.5. Moiré pattern of deflection contours in a long cylindrical brass shell diametrically loaded by a pair of double internal indenters normal to the generators and at the mid-length of the shell.

The same set-up was used for depicting the deflection contours of the deformed matte surface of the loaded shell.

Figure 8.5 shows the moiré pattern of deflection contours

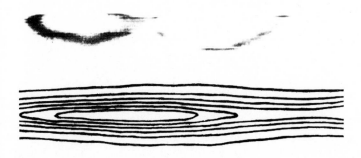

FIG. 8.6. Moiré pattern of deflection contours for a long Perspex cylindrical shell compressed externally by a rubber ring at the mid-length of the cylinder.

created by plastic deformation of a long cylindrical brass shell diametrically loaded by a pair of double internal indenters normal to the generators and at the mid-length of the shell.

Figure 8.6 shows the moiré pattern of deflection contours for a long Perspex cylindrical shell compressed externally by a rubber

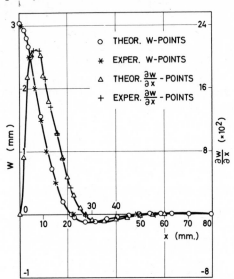

FIG. 8.7. Deflection and slope distributions along a generator of a Perspex cylindrical shell compressed externally by a rubber ring at its mid-length.

ring at the mid-length of the cylinder. The circular cross-section of the rubber ring was of a radius of 0·25 in. and the diameter of the ring was 0·25 in. smaller than that of the shell. The Perspex cylinder with the ring positioned at its mid-length was heated at the rubbery range for Perspex (120°C), where the Perspex cylinder became soft. The set was left to cool to ambient temperature and the ring was then removed. The large elastic deformations frozen in the shell formed the moiré pattern shown in Fig. 8.6. The shell was photographed when unloaded and the original pattern (not shown) contained $1\frac{1}{2}$ initial fringes in the region where the loaded pattern presents the maximum fringe order (Fig. 8.6). Subtraction of the original pattern from the loaded pattern yielded a moiré pattern presenting a radial symmetry with small discrepancies of the order of 2–5 per cent. From eqn. (8.4) the angle of incidence i and the known pitch p of the reference

FIG. 8.8. Buckling configuration of a Perspex cylindrical shell compressed externally by a rubber ring at its mid-length.

grating, height Δz may be calculated. Figure 8.7 presents the deflection distribution along the generator passing through the maximum order of the moiré pattern. The same problem has been solved theoretically by using a solution given by Timoshenko and Woinowsky-Krieger.[12] The agreement between theoretical and experimental results is remarkable.

By excessive compression of the toroidal ring in the experiment

[12] Timoshenko, S., and Woinowsky-Krieger, S., *Theory of Plates and Shells*, 2nd edn., McGraw-Hill, New York, 1959, p. 471.

shown in Fig. 8.6 the Perspex shell buckled and the moiré pattern obtained is shown in Fig. 8.8.

The sensitivity of the method depends on the fineness of the grating RG and on the angle of incidence i.

If the pitch of RG is selected $p = 0.005$ in. and the angle of incidence $i = 20°$, and if each moiré fringe can be located to within one-fifteenth of the interfringe spacing, then

$$\Delta z = 0.001 \text{ in.}$$

The sensitivity of the method can be increased by increasing the frequency of RG. This increase has a limit depending on the total deflection of the shell.

Generally the OBS method is less sensitive than the RIM method described in the next section, and it is suitable for recording large elastic or plastic deflections.

8.2. The reflected image moiré method for the study of slope contours in shells

The method is suitable for depicting the partial slope contours of a deformed shell.

The apparatus used for this method is shown in Fig. 8.9. Coherent monochromatic light, generated by an oblong yellow sodium light source LS (Fig. 8.9a) illuminated the specularly polished lateral surface of the shell. The width and the distance of the light source from the nearest generator of the cylindrical shell were selected such that the narrow illuminated strip complied with the condition of considering this strip as a flat surface. A reference grating RG_1 was placed between the light source LS and the specimen Sp. The lines of the grating were oriented parallel or normal to the generators of the specimen. The pitch p_1 and the distance l_1 of the grating from the nearest generator of the cylinder were selected such as to produce the image RG_1' of the grating RG_1, at a distance l_1' behind the surface, with a suitable frequency. A camera L_1, placed at a distance l_2 from the nearest point of the specimen and at an angle 2ϑ to the incident light beam, received the reflected image of the grating RG_1 on its ground-glass screen.

A second grating RG_2 was juxtaposed to the ground-glass screen of the camera. The lines of this grating were set parallel to the lines of the grating RG_1. The distance $(l'_1 + l_2)$ between the lens L_1 and

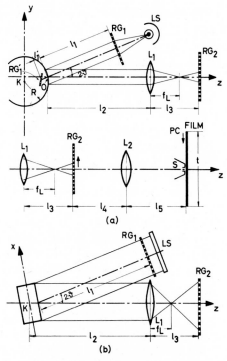

FIG. 8.9. Apparatus used in reflected moiré method.

the image RG'_1, and l_3, between L_1 and RG_2, were adjusted so that the image RG'_1 of the first grating became of equal frequency with the grating RG_2. Interference of the gratings with the specimen unloaded produced along the illuminated generator a uniformly bright field on the ground-glass screen of the camera. The periphery camera PC was placed behind RG_2 and at a distance l_4. This camera photographed the pattern formed on RG_2.

By synchronizing the rotation of the specimen so that the translation of its image on the grating RG_2 is in step with the traverse of the film in the camera PC, a sharp photograph of the development of surface of Sp appears. If the specimen is accurately centred, interference of RG_1' with RG_2 presents a uniformly illuminated field, which, when photographed, shows the correct adjustment of the apparatus.

Figure 8.9a shows the apparatus required for obtaining the moiré patterns of the partial-slope contours normal to the axis of the cylindrical shell, and Fig. 8.9b shows the same arrangement for the case of slope contours parallel to the axis of the cylinder. The only difference between the two set-ups is that the gratings RG_1 and RG_2 were turned through an angle of $90°$ and grating RG_1 has the same frequency as RG_2.

The method may be also applied to non-axisymmetric shells. In this case, if the difference in the extreme radii of the shell is such that the image of RG_1 remains sharp and in focus when it interferes with RG_2, a first photograph with the specimen unloaded yields the geometric shape of the shell and this, when subtracted from the loaded pattern, yields the moiré pattern of partial slopes for the shell due to loading. When the difference between the extreme radii are large the image of RG_1 becomes blurred in regions and out of focus and a technique of photographing the lateral surface of the shell by regions should be adopted. In this case, the beauty of the method due to its simplicity is partially lost.

The method is suitable for measuring the initial macroscopic curvature of a shell as well as small departures due to loading.

In the case where the method is used for the evaluation of the initial macroscopic radius of curvature of the shell R, the focal length of lens L_1, the pitches p_1 and p_2 of RG_1 and RG_2 and the distance l_1, l_2 and l_3 must be measured to within a hundredth of a millimetre with a vernier system. The initial radius of curvature of the shell may be readily and accurately evaluated by using known formulae of elementary optics.

If R is the radius of curvature of the shell at point O (Fig. 8.9a),

l_1 and l'_1 are the distances of the grating RG_1 and its image RG'_1 on the reflecting surface of the shell at O, and p_1 and p'_1 are the pitches of the RG_1 and its image respectively, then from the simple formulae of optics[13] holding for lenses and mirrors it is valid that

$$l'_1 = \frac{p'_1}{p_1} l_1 \tag{8.9a}$$

and

$$R = \frac{2l_1 l'_1}{l_1 - l'_1}. \tag{8.9b}$$

Similarly, for the lens L_1 of the apparatus with a focal length equal to f_L and a grating RG_2 of a pitch p_2 placed at distance l_3 from lens L_1 so that the interference of the image of RG'_1 (with pitch p'_1) and RG_2 does not form moiré fringes, it is valid that

$$\frac{p'_1}{p_2} = \frac{l'_1 + l_2}{l_3} \tag{8.9c}$$

and

$$\frac{1}{l'_1 + l_2} + \frac{1}{l_3} = \frac{1}{f_L}. \tag{8.9d}$$

Solving relations (8.9c) and (8.9d) with respect to p'_1, it is found that

$$p'_1 = \frac{p_1 p_2 (l_2 - f_L)}{(p_1 f_L - p_2 l_1)}. \tag{8.9e}$$

Introducing relations (8.9a) and (8.9e) into eqn. (8.9b) it can be deduced that

$$R = \frac{2l_1(l_2 - f_L)}{\lambda f_L - (l_1 + l_2) + f_L}, \tag{8.10a}$$

where λ expresses the ratio of pitches p_1 and p_2 of the gratings RG_1 and RG_2.

In the case where, instead of the focal length f_L, the distance l_3

[13] Jenkins, F. A., and White, H. E., *Fundamentals of Optics*, 3rd edn., McGraw-Hill, New York, 1957, pp. 28–51.

is measured, it can be readily deduced from the same simple relations (8.9a) to (8.9d) that

$$R = \frac{2l_1 l_2}{\lambda l_3 - (l_1 + l_2)}.$$ (8.10b)

If the distance l_1 is changed, the frequency of the image RG_1' is changed and the interference of RG_1' with RG_2 shows moiré fringes. If the curvature of the cylinder is constant the moiré pattern contains equidistant fringes. Any variation of the curvature will appear with an uneven spacing of the moiré fringes.

If either the focal length f_L of lens L_1 is known and distance l_1 and l_2 are measured, or all three distance l_1, l_2 and l_3 are measured, curvature R of the cylinder can be derived from relations (8.10a) or (8.10b). In the case where no moiré fringes appear from the interference of the image of RG_1 and RG_2, ratio λ may be determined from the fact that the pitch of the image of RG_1' at RG_2 is equal to the pitch of RG_2. If the moiré pattern shows equidistant fringes, ratio λ can be readily derived from the pitch of RG_2 and the interfringe spacing of the moiré pattern. If the moiré fringes show an uneven spacing, λ is variable and curvature R and its variation can be evaluated from the variation of λ. The accuracy in measuring R depends on the accuracy in defining distances l_1, l_2 and l_3 as well as f_L. If vernier systems measure the distance l_1, l_2 and l_3 accurately a higher accuracy can be achieved. With simple measuring devices the overall accuracy in measuring R can be higher than 0·001 in.

Figure 8.10 exhibits the loaded pattern of slope contours along a generator of the same type of loading as that shown in Fig. 8.6 by applying the OBS method. The deformed shell remains axially symmetric after loading and there is no moiré pattern in a direction normal to the axis of the cylinder. The one pattern suffices for the evaluation of the slope distribution. Figure 8.7 presents the slope distribution along a generator obtained after subtracting the original moiré pattern with the shell unloaded from the loaded moiré pattern in Fig. 8.10. The experimental results compared well with the theoretical values of slopes.

The incident rays with the shell deformed will deflect by an angle $(\vartheta + 2\varphi)$ in the Oyz plane because the normal of each strip of the lateral surface of the shell entering into the illumination region of the light source is forming an angle $(\vartheta + \varphi)$ with the incident ray of light, due to a deflection w of the shell. The fringe pattern formed from the interference of the reflections of grating RG_1 and grating

FIG. 8.10. Moiré pattern of slope contours along a generator of a long Perspex cylindrical shell compressed externally by a rubber ring at its mid-length.

RG_2 yields the partial slope contours in a direction normal to the lines of the gratings.

From elementary optics angle φ is expressed by

$$\varphi = \frac{N p_1 \cos^2 \vartheta}{2 l_1}, \tag{8.11}$$

where N is the fringe order corresponding to the variation φ of the angle of incidence ϑ at point O, and l_1 is the distance between RG_1 (of pitch p_1) and the illuminated strip of the lateral surface of Sp.

The fringe order N of the moiré pattern is proportional to angle φ and therefore proportional to slope $\partial w / \partial n$ in the direction n, normal to the rulings.

Therefore

$$N = C\frac{\partial w}{\partial n}, \tag{8.12}$$

where C is a factor of proportionality depending on the geometry of the apparatus and the relation $\lambda = p_1/p_2$ of the pitches of the two gratings.

Since slopes and curvatures are functions depending on direction, an orthogonal set of slope or curvature contours are necessary through each point for the complete definition of the surface. Hence, two moiré patterns formed from gratings, which have their lines parallel to the Ox and Oy axes respectively, are necessary and sufficient for the complete definition of the shell surface. With the lines of the gratings parallel to the axes of symmetry of the specimen, each of the moiré patterns yields the partial slope distribution along one of the axes. Along any other traverse of the specimen both contour patterns contribute to the evaluation of the slope distribution along the traverse and the partial slopes in a direction parallel or normal to a traverse forming an angle ω with x-axis are given by relations (6.17).

A numerical or graphical differentiation of the partial slope distribution along any direction yields the curvature along the same line. For the evaluation of the moment and strain distribution in a cylindrical shell elastically loaded it suffices to evaluate the partial curvatures $\partial^2 w/\partial x^2$ and $\partial^2 w/\partial y^2$, as well as the twist $\partial^2 w/\partial x\partial y$ for each point of the shell and to determine its flexural rigidity. The flexural rigidity may be determined from the mechanical properties of the material and the thickness of the shell.

For the evaluation of the proportionality factor C it is necessary to calibrate the moiré pattern. The fringe order of the resultant moiré pattern after subtracting the unloaded pattern depends on the geometry of the optical apparatus and especially on angle ϑ, the principal distances l_1, l_2 and l_3 and the focal length f_L, as well as on the pitches p_1 and p_2 of the two gratings RG_1 and RG_2. In order to check the value of the constant C the maximum and minimum diametral deflections of the shell were measured by a sensitive mechanical gauge and compared with the values of deflections derived from integration of the appropriate slope-contour patterns.

The sensitivity of the method depends on the fineness of the gratings and on the distance between the shell specimen and the reference grating RG_1. Since in this method the same set-up was used as utilized in § 6.6 for the recording of partial slope contours in flexed plates the sensitivity remains the same and takes the value of $2 \cdot 5 \times 10^{-5}$ rad per fringe. The sensitivity of $2 \cdot 5 \times 10^{-5}$ rad is very satisfactory and small deflections of shells can be determined by the moiré fringes.

8.3. Hologrammetry used for recording partial slopes of plates and shells

The wavefront reconstruction technique was first proposed by Gabor[14] in 1948 and improved by Leith and Upatnieks[15] to the point where space images of three-dimensional bodies can be recorded. Horman[16] suggested the use of the wavefront reconstruction technique for recording transient phenomena which, superposed, allowed the performance of interferometric measurements. Powell and Stetson[17] utilized the same technique for vibration measurements. Finally, Haines and Hildebrand[18] used the interference phenomena observed between the deformed object and its hologram taken at the unloaded state to determine the deflections, translations and rotations which a certain point of the deformed body has undergone. The theory of the wavefront reconstruction

[14] Gabor, D., *Nature* **161**, 777 (1948); *Proc. Roy. Soc. London* **A197**, 454 (1949), and **B64**, 449 (1951).

[15] Leith, E., and Upatnieks, J., *J. Opt. Soc. Am.* **52** (10) 1123 (1962); *J. Opt. Soc. Am.* **53** (12) 1377 (1963), and *J. Opt. Soc. Am.* **54** (11) 1295 (1964).

[16] Horman, M. H., *Appl. Opt.* **4**, 333 (1965).

[17] Powell, R. L., and Stetson, K. A., *J. Opt. Soc. Am.* **55**, 1593 (1965).

[18] Haines, K. A., and Hildebrand, B. P., *Appl. Opt.* **5** (4) 595 (1966); *Appl. Opt.* **4**, 172 (1965).

technique will be concisely presented following the development by Haines and Hildebrand.

Wavefront reconstruction is a technique for recording specific data about a three-dimensional body on a photographic film and then reconstructing a three-dimensional image of the object from this recording. The reconstruction of the three-dimensional image is achieved by recording the wave pattern of the light scattered by the body instead of recording the image of the body in an ordinary camera. For this purpose the light used must be coherent in space and time. The most convenient source of a strongly coherent light is a laser.

The recorded wave pattern of the scattered light from the object is called a *hologram*. In the recording process the phase of the incident illumination is lost, but it was shown by Gabor that for a strong coherent background the loss of phase is less important and a fairly good image of the original body can be recovered from the intensity record alone. It was shown by Leith and Upatnieks that a hologram made from a diffusely illuminated object yields a reconstruction free of flows resulting from dust particles, etc., in the optical system. Moreover, the two-beam technique adapted to three-dimensional bodies creates reconstructions resembling to a high degree the original objects. The light produced by a laser is highly coherent and monochromatic. While these features make the laser an excellent light for hologram reconstructions of high quality, hologram reconstructions can be also made using a conventional mercury arc lamp in combination with an interferometer. This technique appears to work equally well as the method using a laser but is more difficult to carry out experimentally. In some way the moiré patterns of partial slope contours obtained by the reflected image moiré method either in two-dimensional problems or in problems of flexed plates may be considered as formed by a similar procedure as the procedure of holography. The specimen grating RG_2 represents the image of the undeformed body which interferes with the image of grating RG_1 distorted by the deformation of the body.

Consider a beam of coherent monochromatic light which

illuminates the specimen and the light reflected from it falls on a photosensitive surface. Let the light amplitude falling on the film be expressed as

$$s(x, y) = a(x, y) \cos [\omega t + \varphi(x, y)], \qquad (8.13)$$

where the system of Cartesian coordinates x, y coincides with the plane of the film. Frequency ω is the radian frequency of light and $a(x,y)$ and $\varphi(x,y)$ are the amplitude and phase modulations imposed on the light beam by the object.

A part of the light beam, called the *reference beam*, is intercepted by a mirror and directed on to the film without any disturbance. The reference beam may be expressed as

$$f(x, y) = \alpha_r(x, y) \cos [\omega t + \alpha x + \beta r^2], \qquad (8.14)$$

where $\alpha = (2\pi/\lambda) \sin \xi$ and $\beta = \pi/(\lambda z)$ for $z \gg x$. Here ξ is the average angle formed by the reference beam and the (x,y) plane of the film, z the distance between the light source and the plane of the film, and r the polar distance on the x, y plane ($r^2 = x^2 + y^2$). The plane of the reference source is coinciding with the xz plane and β represents the curvature of the wavefront which depends on distance z of the point source to the film plane.

The reference beam and the specimen beam add on the plane of the film and create a total amplitude $A(x,y)$ given by

$$A(x, y) = a(x, y) \cos [\omega t + \varphi(x, y)] + a_r(x, y) \cos [\omega t + \alpha x + \beta r^2].$$
$$(8.15)$$

Since the film is an energy detector, only the intensity of the wave can be recorded. Then

$$|A(x, y)|^2 = |s(x, y)|^2 + |f(x, y)|^2 + (fs)_+ + (fs)_-, \qquad (8.16)$$

where

$$\left. \begin{aligned}
|s(x, y)|^2 &= \tfrac{1}{2}|a(x, y)|^2 . \{1 + \cos 2[\omega t + \varphi(x, y)]\}, \\
|f(x, y)|^2 &= \tfrac{1}{2}|a_r(x, y)|^2 . \{1 + \cos 2[\omega t + \alpha x + \beta r^2]\}, \\
(fs)_+ &= \tfrac{1}{2}a(x, y)a_r(x, y) \cos [2\omega t + \alpha x + \beta r^2 + \varphi(x, y)], \\
(fs)_- &= \tfrac{1}{2}a(x, y)a_r(x, y) \cos [\alpha x + \beta r^2 - \varphi(x, y)].
\end{aligned} \right\} \qquad (8.17)$$

The terms in relations (8.17), which contain the radian frequency, average to zero since the photographic process is a time-averaging process. Then the only terms which remain from relations (8.17) are the terms

$$\tfrac{1}{2}|a(x, y)|^2, \quad \tfrac{1}{2}|a_r(x, y)|^2 \quad \text{and} \quad (fs)_-.$$

If the exposed film is developed and then placed in the same position it occupied during the exposure procedure, while the specimen is removed from its place, the illumination on the film remains the same as this expressed by eqn. (8.14). The light transmitted from the film is the product of the impinging wave and the transmission pattern on the film. Since the terms

$$\tfrac{1}{2}|a(x, y)|^2 \quad \text{and} \quad \tfrac{1}{2}|a_r(x, y)|^2$$

do not contribute to the process, only term $(fs)_-$ is important and the transmitted light amplitude is expressed as

$$f(x, y)[(fs)_-] = \tfrac{1}{2}a(x, y)a_r^2(x, y) \cos\left[\alpha x + \beta r^2 - \varphi(x, y)\right]$$
$$\cos\left[\omega t + \alpha x + \beta r^2\right]. \quad (8.18)$$

Expression (8.18), expanded, yields

$$f(x, y)[(fs)_-] = \tfrac{1}{4}a(x, y)a_r^2(x, y)\{\cos\left[\omega t + 2\alpha x + 2\beta r^2 - \varphi(x, y)\right]$$
$$+ \cos\left[\omega t + \varphi(x, y)\right]\}. \quad (8.19)$$

Relation (8.19) shows that the second term, except for an amplitude weighing factor $a_r^2(x, y)$, is a reproduction of the wave generated by the specimen when it was recorded. Hence this term reconstructs the object in exact detail including all the effects of normal three-dimensional viewing.

The first term in expression (8.19) represents a wave leaving the film surface at an angle proportional to 2α with a curvature 2β. By choosing ξ sufficiently large the light from the first term diffracts out of the field of view of the second term leaving only the contribution of the second term. In this case the first term can be ignored.

The technique described above results in the exact duplication of the wavefront reflected from the specimen. The wavefront may be used in an interferometer to obtain a complete picture of the deformed specimen due to application of an external load. The reconstructed wavefront is used as the reference against which to compare the wavefront reflected by the strained specimen. The apparatus used for this process is shown in Fig. 8.11.

A gas laser was used as a light source. A beam splitter divided the laser beam into a reference beam and a specimen beam. The

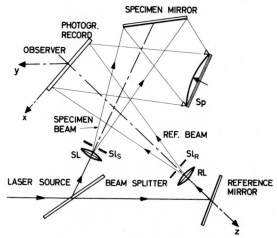

FIG. 8.11. Apparatus used for wavefront reconstruction of a strained specimen by the two-beam technique.

reference beam, reflected from the front surface of a mirror, was normally incident to the photographic plate after passing through a lens and a pinhole. The specimen beam, after reflection from the beam splitter, became divergent by passing through a lens and a pinhole and it was reflected from the front surface of a mirror and illuminated the specimen. The light scattered from the specimen impinged on the surface of the photographic plate which was also exposed to this light. This arrangement is the so-called *two-beam technique*.

In the wavefront reconstruction the object was removed after taking the photograph. The hologram, which is then exposed in the double beam and developed plate, takes the place of the initial photographic plate and is viewed by the same light source. In this case a three-dimensional image of the specimen appears at the initial position of the missing object.

If the object is not removed from its position and the hologram is placed at the position of the photographic plate, the wavefront reflected from the object interferes with the wavefront generated by the hologram producing interference fringes. If the film plane

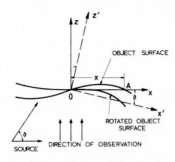

FIG. 8.12. Schematic representation of the geometry of a flexed or purely rotated specimen.

is adjusted very carefully, an exact coincidence between the object and image is achieved and the fringes disappear. If the object is submitted to a deformation the wavefront of the deformed object interferes with the wavefront of the undeformed hologram producing an interferogram due to straining.

Consider the simple case where the fringes are formed on the object. This is the case where flexed and laterally deformed bodies are considered. Another case is where the objects are submitted to a pure rotation. In this case the coordinate system is attached to the point on the object about which the surface is rotated. If a generic area of a surface is rotated through an angle β due to the overall deflection or rotation of the body (Fig. 8.12) the path difference S of the light from the source to the observer via the object and corresponding image points at point A, which is at a distance x from the origin O, is given by

$$S = x\beta(1 + \sin \vartheta), \qquad (8.20)$$

where ϑ is the angle subtended by the object surface and the incident light beam. The radiation from the object and the image differs only by a linear phase φ which, according to the approximate theory mentioned above, is given by

$$\varphi = \left(\frac{2\pi}{\lambda}\right) x\beta(1 + \sin \vartheta). \qquad (8.21)$$

A system of parallel fringes is formed at the vicinity of A due to the rotation of the body which lies on the surface of the specimen. The orientation of the fringes is parallel to the axis of rotation (which in this case is the y-axis).

The interfringe spacing f_M at the vicinity of the point A is given by

$$f_M = \frac{\lambda}{\beta}(1 + \sin \vartheta)^{-1}. \qquad (8.22)$$

The angle of rotation β may be found by measuring the interfringe spacing f_M and angle ϑ.

The interferogram shows the lateral displacement contours of the deformed or rotated body. Figure 8.13 shows the effect of the flexure of a circular plate flexed by a concentrated load applied at the centre of the plate. Figure 8.13a shows a slight deflection of the plate (1 μ), while Fig. 8.13b represents a larger deflection (5 μ).

The same method may be applied to the study of deformation of a body subjected to any strain field. In this case the interference fringes are formed at a distance of at least several thousand wavelengths from the object surface. The analysis for this type of deformation is quite complicated and is beyond the scope of this study.

Hologrammetry, as it was sketched in this section, is a very useful measuring technique, although its potentialities have not yet been fully explored as the method is still at its infancy. It belongs to the category of optical interferometric methods and is therefore beyond the realm of moiré methods. However, it is hoped that the

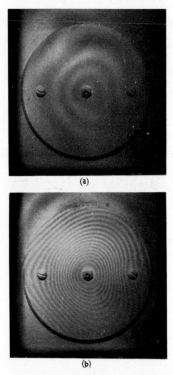

(a)

(b)

FIG. 8.13. Interferograms of slope contours obtained by the two-beam technique of a metallic circular plate flexed by a concentrated central load. (a) For an overall deflection of the plate of 1 μ. (b) For an overall deflection of the plate of 5 μ. (*Courtesy of Dr. K. A. Haines.*)

method may be extended to coarse-grating moiré techniques yielding a potential tool for the study of the state of deformation of complicated three-dimensional bodies. The purpose of this section is therefore to stimulate experimenters to extend the already known moiré techniques to this fruitful domain.

CHAPTER 9

Moiré Extensometers

THE idea of using the moiré phenomenon as a measuring device in extensometers is very attractive and researchers have utilized it in producing various types of extensometers. Apparently the first who used gratings as elements of extensometers were Wiemer, Lehmann and Voigt.[1] Crisp,[2] in 1957, introduced an instrument for measuring the interfringe spacing as well as the relative angular displacements of the two gratings and of the fringe pattern. Linge,[3] again in 1957, used the interference of two crossed gratings for the measurement of lengths. Diruy[4] introduced an extensometer appropriate either for the measurement of static deformations (version A) or dynamic deformations (version B). In both cases angular disparity moiré fringes were utilized for evaluating the elongations over a rather large gauge length.

Finally, Holister, Jones and Luxmoore[5] developed an instrument where the measuring device was based on the displacement of vernier moiré fringes.

9.1. Wiemer's moiré extensometer

The vernier moiré effect was used by Wiemer, Lehmann and Voigt[1] as a means of measuring relative displacements in combination with a mechanical magnification system.

[1] Wiemer, A., Lehmann, R., and Voigt, H., *Feingeräte Technik* **3** (4) 161 (1954).

[2] Crisp, J. D. C., *Proc. Soc. Exp. Stress Anal.* **15** (1) 65 (1957).

[3] Linge, J. R., *Aircr. Engng.* **29** (337) 70 (1957).

[4] Diruy, M., Laboratoire Central des Ponts et Chaussées, Report 63–7 (1963).

[5] Holister, G. S., Jones, W. E. M., and Luxmoore, A. R., *Strain* **2** (4) 27 (1966).

Figure 9.1 illustrates diagrammatically the operation of the instrument. A fixed contact knife edge *a* was attached to frame *m* of the instrument, while the rotating knife edge *b* was integral with arm *h* and could rotate in a flexure pivot attached to the frame. Its rotation resulting from the elongation of the specimen magnified

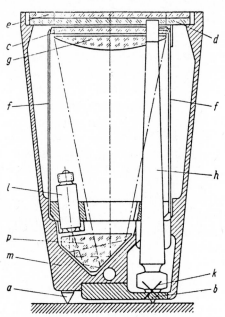

FIG. 9.1. Schematic diagram of
Wiemer's moiré extensometer.

and transmitted it to the sliding grating *c*, which was connected to arm *h* through a joint free of bending. Sliding grating *c* was guided by a flat spring which drew the sliding grating back to its initial position. There were lateral sleeves on the frame which guided the sliding grating to slide always parallel to itself. A stationary grating *d* was superposed on the sliding grating and adjusted to fixed sleeves so that the moiré pattern formed by the superposition of the two gratings corresponded only to a linear disparity between

he gratings. Contribution to the moiré pattern from angular disparity was in this way completely excluded. On the top of the two
gratings a matte glass plate was housed on grooves of the frame.
A microscale was traced on the matte glass plate.

The illumination of the gratings was achieved by a white-light
lamp l, the divergent light beam of which was totally reflected by
the prism p and impinged on a planoconvex lens g placed underneath the sliding grating. With the light being switched on, the
illumination of the gratings was sufficient to make the moiré
pattern visible at normal ambient illumination.

The gauge length of the instrument was small and equal to
10 mm, and the smallest elongation which was claimed to have
been measured by this instrument was 1μ . For a distance between
two successive marks on the scale equal to 2 mm, the magnification
was approximately equal to 2000. This magnification was divided
into two parts, (a) the mechanical magnification ($\times 20$) achieved
by the lever formed by the rotating knife edge and arm h, and (b)
the optical moiré magnification ($\times 100$). In order to obtain a moiré
magnification equal to 100 the pitches of the two gratings were
taken as $p_s = 0{\cdot}202$ mm and $p_r = 0{\cdot}200$ mm. The interfringe spacing
for these gratings was, therefore,

$$f = \frac{p_s p_r}{p_s - p_r} = 20{\cdot}2 \text{ mm}, \qquad (9.1)$$

and the moiré magnification M was

$$M = \frac{p_s}{p_s - p_r} = 101 \approx 100. \qquad (9.2)$$

The glass plates containing the fine gratings were divided into
four narrow strips, the one extreme strip being occupied by the fine
gratings described above with pitches approximately equal to
$0{\cdot}200$ mm. The central strip contained an optical indicator of the
sign of measured deformation (elongation–contraction), while the
remaining two strips contained coarse line gratings having a pitch
$p_s = 2$ mm for the gratings on the steady glass plate and a pitch

$p_{cr} = 1 \cdot 8$ mm for the gratings on the sliding glass plate. The one
pair of the corresponding gratings in both plates was displaced
relative to the other pair by half a pitch so that the black strips of
the one pair corresponded to the transparent strips of the other
pair. The moiré magnification for this pair of gratings M_c was
given by

$$M_c = \frac{p_{cs}}{p_{cs} - p_{cr}} = 10, \qquad (9.3)$$

and the new interfringe spacing f_c was

$$f_c = \frac{p_{cs} \cdot p_{cr}}{p_{cs} - p_{cr}} = 18 \text{ mm}. \qquad (9.4)$$

Then the moiré fringes formed by the pairs of the coarse
gratings, while they had an interfringe spacing $f_c = 18$ mm, presented a magnification of 10 and therefore acted as coarse vernier
moving slowly with a changing deformation. A finer estimation of
the measured deformation was made by the fine pair of gratings
with a magnification equal to 100.

The combination of moiré vernier fringes formed by fine and
coarse gratings forms the so-called *multistage vernier moiré system*
and results in a continuous and highly accurate pursuit of the
evolution of deformation.

The instrument was compared with a universal measuring microscope and the agreement of the results was satisfactory, the
maximum scatter being of the order of $\pm 0 \cdot 3 \ \mu$. The strain range
of the instrument is ± 100 divisions, i.e. is 200 μ.

A similar measuring device based on vernier moiré fringes for
the measurement of deformations was recently introduced by
Holister, Jones and Luxmoore.[7]

The extensometer contained a fixed grating mounted on the
frame of the gauge and a similar sliding grating pivoting on two
levers, one of which was the driver arm, giving a magnification

[6] Guild, J., *Diffraction Gratings as Measuring Scales*, Oxford Univ. Press,
London, 1960, p. 198.

[7] Holister, G. S., Jones, W. E. M., and Luxmoore, A. R., *Strain*, **2** (4) 2
(1966).

factor of 10, while the other was an idler arm. Thus the magnification ratio by mechanical means was 10, while the optical magnification due to moiré effect was 100, the total being equal to 1000. The instrument used the sharpening effect by employing complementary line gratings with a transmittance smaller than 50 per cent. It was proved that by an effective sharpening of the moiré fringes achieved by using a reference grating with transmittance 6·25 and a sliding grating with transmittance 1/6·25, a moiré fringe can be located to the one-hundredth of the interfringe spacing.[8] The frequencies of gratings were: $d_r = 100$ lines per inch

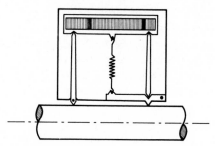

Fig. 9.2. Schematic diagram of Holister's moiré extensometer.

and $d_s = 99$ lines per inch so that the interfringe spacing $f_f = 1$ in. Then a moiré fringe can be located to within $\pm0\cdot01$ in. Therefore the theoretical sensitivity of the gauge is 10^{-5}.

The gratings were illuminated by a miniature light bulb situated at the focus of a parabolic reflector which illuminated a light diffuser. Since there was an air gap between the fixed grating and the sliding grating of the order of 0·005 in., which introduced a parallax error when the fringe pattern was not viewed normally, a sliding cursor was constructed from a Perspex $\frac{1}{4}$ in. thick plate attached to the viewing side of the grating aperture traced with two reading datum lines parallel to the rulings of the gratings. The fringe position reading was taken by aligning the datum lines with

[8] Zandman, F., Holister, G. S., and Brcic, V., *J. Strain Anal.* **1**, 1 (1965).

the estimated centre of the fringe thus ensuring that the viewing direction is approximately perpendicular (Fig. 9.2).

The instrument is lacking the second moiré pattern formed by a pair of coarse gratings with small sensitivity existing in Wiemer's moiré extensometer, which was used as a follower of the moiré fringe movement and excluded any possibility of losing the sequence of evolution of deformation.

9.2. Crisp's moiré goniometer

Crisp[9] introduced an instrument suitable for the measurement of angle ϑ subtended between the rulings of two angularly displaced line gratings and angle φ subtended between the moiré fringes formed by these gratings and one family of rulings. By measuring the interfringe spacing at a generic area of the displacement field, Crisp defined the applied strain to the specimen by an already-traced strain nomograph. He called his device an interferometer, but this instrument is rather a goniometer.

If the pitches of the reference and the specimen gratings are $p(1+\lambda)$ and p respectively, and angles ϑ and φ are defined as above, it is valid that

$$\frac{1}{p^2(1+\lambda)^2} - \frac{2\cos\vartheta}{p^2(1+\lambda)} + \frac{1}{p^2} = \frac{1}{f^2}, \qquad (9.5)$$

where f is the interfringe spacing.

Similarly, the following relation holds in the case where the angle φ is involved:

$$\frac{1}{p^2(1+\lambda)^2} - \frac{2\cos\varphi}{pf(1+\lambda)} + \frac{1}{f^2} = \frac{1}{p^2}. \qquad (9.6)$$

Relations (9.5) and (9.6) indicate that the quantities $1/p$, $1/p(1+\lambda)$ and $1/f$ can be represented by the three sides of a triangle, which has as angles the angles φ and ϑ (Fig. 9.3).

For solving eqn. (9.6) a nomograph can be constructed where a system of radial lines and circular arcs is drawn using as centres

[9] Crisp, J. D. C., *Proc. Soc. Exp. Stress Anal.* **15** (1) 65 (1957).

the ends of a line of unit length and then a set of such triangles may be formed on this graph. By interpolation, the apex of any required triangle given any two of the parameters ϑ, φ and p/f can be found and hence the pitch of the strained specimen grating $p(1 + \lambda)$ can be evaluated. Such a nomograph is given by Crisp, which is drawn to yield a direct strain reading since the strain

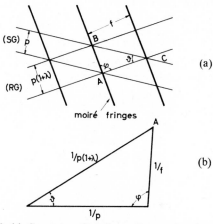

FIG. 9.3. (a) Geometry of grating rulings and moiré fringes for an angular displacement between the two gratings. (b) Strain triangle relating grating characteristics and the interfringe spacing.

measured on a gauge length equal to the reference pitch p is given as $\varepsilon = \lambda$.

For the required accurate measurement of angles ϑ and φ and interfringe spacing f, an instrument was devised by Crisp which also incorporated the reference grating. The instrument essentially consisted of a main frame holding two rotating rings. The first ring supported the reference grating, which was reproduced on a polystyrene base film for reducing the thermal and humidity expansions or contractions of the film. The second ring supported a piece of celluloid plate bearing a scribed target with lines in orthogonal directions for the purpose of alignment with the observed fringes and evaluation of angle ϑ, as well as of estimation of interfringe spacing f. The master grating and the scribed celluloid disc were placed in close proximity to avoid parallax, and the frame was so arranged that the specimen and reference gratings were in contact. Scales of angles were engraved on the two rotating

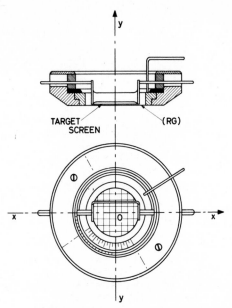

Fig. 9.4. Plan and section schematic diagram of Crisp's goniometer.

rings, which enabled the measurements of the relative angular displacement of the reference grating and the moiré fringes (angles ϑ and φ respectively). Figure 9.4 shows a schematic representation of the plan view as well as of a section of the instrument.

The gratings used for the goniometer were of a frequency of 100 lines per inch, and the accuracy of measurements was claimed to be very high. The errors in strain reading were of the order of 0·005–0·008 for suitable type of measurement and magnitudes of strain (larger than 5 per cent). With the use of a finer grating the error may be further reduced. Moreover, a further improvement in accuracy may be achieved by tracing finer protractor scales.

9.3. Moiré extensometers based on angular disparities

In this type of extensometers elongations are related to angular

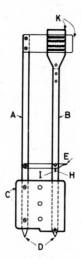

Fig. 9.5. Schematic diagram of Linge's extensometer.

disparities of gratings of equal pitch. The types of extensometers produced by Linge and Diruy will be described.

9.3.1. *Linge's moiré extensometer*

A schematic diagram of the extensometer is shown in Fig. 9.5. The extensometer essentially consisted of two bars of tool steel *A* and *B*, which were sandwiched between two metal sideplates *C*. While the bar *A* was bolted to plates *C*, bar *B* was free to rotate on integral edges in the V-notches cut into the sideplates.

Both bars terminated in chisel points *D* at their lower ends and two line gratings K were attached to the other extremities of the bars, suitably oriented to yield moiré fringes whose direction was normal to the longitudinal axis of the extensometer. A locking bar E loaded by the return spring I prevented the movement of bar B, while the extensometer was attached to the surface under test; the bar was held in the test position by spring H.

Strain experienced by the chisel points D was transmitted by the rotating bar B to cause relative movement between the two gratings and a shift of the moiré fringes initially observed.

Basic considerations show that the number of fringes which pass a reference dot marked on the grating attached to bar B yields a measure of the deformation applied between chisels D.

The grating attached to the moving arm B could be angularly displaced relatively to the bar for increasing or decreasing the interfringe spacing of the moiré pattern. The angular adjustment of the grating causes a small fringe shift which, in addition to the change in width, could be used to set the zero on any given fringe thereby avoiding the necessity of estimating parts of fringes at the outset.

The gratings used were of a frequency of 500 lines per inch (20 lines per millimetre) and the mechanical magnification of the instrument was 23·32. Simple calculations show that the movement of one fringe at the head of the extensometer corresponded to an elongation $\Delta l = 0.0000844$ in. between the chisel points.

The sensitivity of the instrument was variable since the angular displacement of the two gratings may be split into two linear components, the one normal to the rulings, which does not contribute to the displacement of moiré fringes, and the other parallel to the rulings, which causes the displacement of the moiré pattern. The ratio between maximum and minimum sensitivities was estimated by the constructor to be of the order of 5.

9.3.2. *Diruy's moiré extensometer*

The characteristics of Diruy's extensometer are: gauge length 5·9 in. (150 mm), measuring range of 150 μ, sensitivity of 0·05 μ. It is entirely free from time drift and it is equipped with a compensation device which makes the instrument practically insensitive to variations of room temperature.

Two types of extensometers were produced: type A, suitable for static or slowly varying deformations, and type B, suitable for dynamic measurements.

Diruy's type A extensometer is shown in Figs. 9.6 and 9.7. It essentially contained two parallel prismatic bars P_1 and P_2, which were connected at their extremities by thin and flexible elastic springs. A small relative displacement between the two bars was

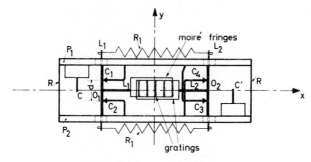

FIG. 9.6. Schematic diagram of Diruy's extensometer (type A).

possible along the longitudinal axis xx of the instrument by a small flexure of the flat springs, while the relative displacement along a transverse axis yy was insignificant.

Two knife-edges C and C', integral with P_1 and P_2 respectively, were used to define the gauge length of the instrument and to fasten it to the testpiece. Distance CC' constituted the gauge length l. Every variation of the gauge length Δl provoked a relative displacement of the bars. C_1, C_2, C_3, C_4 were knife-edges. The first two were integral with bar P_1 and the latter two with the bar P_2. The pointers of the edges were vertical and they are schematically presented by arrows in Fig. 9.6. Distances C_1C_2 and C_3C_4 separating the pointers of the knife-edges were equal and are designated as $C_1C_2 = C_3C_4 = d$.

Two levers L_1 and L_2 in the form of T were connected with each pair of knife-edges respectively, i.e. lever L_1 together with C_1 and C_2 and lever L_2 with C_3 and C_4. Two springs R_1 fixed between L_1

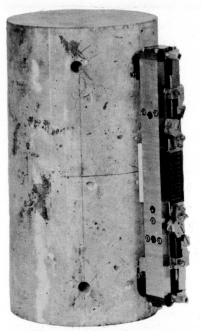

Fig. 9.7. Diruy's extenso-meter (type A) mounted on a concrete cylindrical speci-men. (*Courtesy of Mr. M. Diruy.*)

and L_2 assured a satisfactory application of the levers on their respective knife-edges.

A line grating of pitch $p = 0.050$ mm reproduced on a photo-graphic plate was attached to the extremity of each lever system. The planes of the two gratings were parallel, nearly in contact between them, and perpendicular to the axes of rotation of the levers (O_1 and O_2). The directions of their rulings formed a small angle α the bisector of which was parallel to the longitudinal axis of the instrument.

A relative displacement of the bars P_1 and P_2 was transformed into two equal angular displacements of opposite sign of the two levers to which the gratings were attached. This transformation of a linear displacement into an angular displacement was effectuated by the pairs of knife-edges C_1, C_2 and C_3, C_4. A variation Δl of the distance CC' produced a displacement Δl of C_2 with respect to C_1 and of C_4 with respect to C_3 in the direction xx. Lever L_1 was therefore angularly displaced by an angle φ, while the lever L_2 was displaced by an angle $-\varphi$ and the variation of distance l was given, for small angles φ, by

$$\Delta l = d \cdot \tan \varphi \approx d \cdot \varphi. \tag{9.7}$$

The direction of rulings in the two gratings changed and became $(\alpha + 2\varphi)$ and the initial interfringe spacing f which is given by

$$f = \frac{p}{2 \sin (\alpha/2)} \approx \frac{p}{\alpha}, \tag{9.8}$$

became

$$f' = \frac{p}{2 \sin (\alpha + 2\varphi)/2} \approx \frac{p}{\alpha + 2\varphi}. \tag{9.9}$$

Introducing relations (9.8) and (9.9) into (9.7) it results that

$$\Delta l = \frac{pd}{2}\left(\frac{1}{f'} - \frac{1}{f}\right). \tag{9.10}$$

Relation (9.10) implies that, by measuring the initial and the final interfringe spacings f and f', the deformation Δl may be evaluated.

The sensitivity of measurement of the interfringe spacings allows for the measurement of a deformation equal or inferior than 150 μ to an approximation equal to 4×10^{-2} μ. The sensitivity of the instrument is constant all over the range of measurements of the instrument.

The reproducibility of the measurements may be achieved to an approximation of $5 \times 10^{-2}\,\mu$. The instrument does not present any kind of time drift.

The use of extensometer type A was reserved for static or slowly variable deformations. Type B was suitable for dynamic measurements. The difference between instruments of type A and type B, which modified completely the principle of measuring, was that the positions of knife-edges C_3 and C_4 were permutated (Fig. 9.8). Their pointers remained at their positions but knife-edge C_3 was integral with P_2 and C_4 with P_1 respectively. Consequently, levers

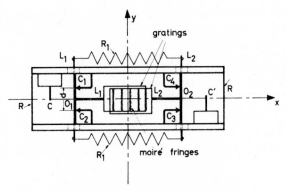

Fig. 9.8. Schematic diagram of Diruy's extensometer (type B).

L_1 and L_2 rotated in the same direction and the angular difference α between the rulings of the gratings remained constant during the deformation. Therefore the interfringe spacing f remained constant and the fringes were translated in the direction xx which coincided with the direction of the bisector of angle α between the rulings of the gratings.

If each grating was angularly displaced by an angle φ it is valid that

$$\frac{\Delta l}{d} = \tan \varphi \approx \varphi. \tag{9.11}$$

Moreover, the relative translation of the two gratings in the transverse direction yy is given by

$$\Delta y = \varphi l', \tag{9.12}$$

where l' is distance $O_1 O_2$, which corresponds to the passage of $(n + \lambda)$ fringes before the reticle of the telescope of the instrument. While n expresses the integral orders of fringes that have passed before the reticle, λ is expressing the fractional excess of the last fringe $(0 \leqslant \lambda < 1)$.

It is valid that

$$\Delta y = p(n + \lambda) \tag{9.13}$$

and

$$\Delta l = d\varphi = d\frac{\Delta y}{l'}. \tag{9.14}$$

Therefore

$$\Delta l = \frac{(n + \lambda)pd}{l'}. \tag{9.15}$$

Since the instrument was designed to measure transient strains the use of an aiming telescope to count the passage of the moiré fringes is rather inconvenient. It was therefore preferable to use a photoelectric means for recording the moiré fringes. For this purpose the instrument was related with photoelectric cells, the signals of which, after amplification, were recorded by a recorder.

Diruy indicated that by fastening one of the prismatic bars to the vertical column of a rigid stand the instrument could be used in both its versions as high sensitivity comparator.

9.3.3. Vafiadakis' moiré extensometer

The simple displacement gauge shown in Fig. 9.9 capable of operating submerged in a hot oil bath was fabricated utilizing the principle of relative angular displacement of gratings. This gauge was recently introduced by Vafiadakis.[10] The gauge utilized two identical gratings, one being attached to the body of the instrument and the second to an arm pivoting about a point. The linear displacement v, to be measured, was transmitted to the arm by a probe located at a distance R from the pivot.

[10] Vafiadakis, A. P., *J. Sci. Instrum.* **44** (12) 1008 (1967).

From the geometry of the instrument and the relationship of fringe density along an axis parallel to the grid lines of the pivoting grating N_x it can be shown that for small values of relative angular displacement, magnification is given by

$$\frac{N_x}{v} = \frac{1}{Rp}. \qquad (9.16)$$

FIG. 9.9. Vafiadakis' displacement gauge for high temperatures. (*Courtesy of Dr. A. Vafiadakis.*)

Since in the particular instrument line density $1/p = 500$ and radius $R = 0.5000$ in. $N_x/v = 1000$ fringes per inch per unit displacement. For convenience in the evaluation of results the area of superposition of the grating was chosen as 1 in. long and the width as $\frac{1}{4}$ in. in order to clearly define the fringes. Thus, every addition or reduction of one fringe along the length of the angularly displaced grating (1 in.), which was taken as the scale of the instrument, corresponded to a displacement of 0.001 in. over a range of about 0.100 in.

CHAPTER 10

Precision and
Influence of Grating Defects

10.1. Precision

It has been already mentioned that the precision of measurements by moiré fringes depends on the frequency of the gratings. The highest frequency which can be satisfactorily reproduced by a photographic process is of the order of 1000–1250 lines per inch (40–50 lines per millimetre). This is an upper limit for frequencies used in strain measurement applications of moiré fringes.

It is also known that for the equal pitch method an extension ε creates moiré fringes the interfringe spacing f of which is

$$f = \frac{p}{\varepsilon},\tag{10.1}$$

where p is the pitch of both gratings.

For gratings of a frequency of 1000 lines per inch an extension equal to 10^{-3} in. yields fringes spaced to a distance $f = 1$ in. and an extension equal to 10^{-2} in. gives a moiré pattern with $f = 0 \cdot 1$ in.

If the linear differential moiré method is used the sensitivity of the method is considerably increased. If the initial interfringe spacing between two gratings of pitches p and $p(1 + \lambda)$ is f, a strain ε applied to one of the gratings will change the interfringe spacing to $f' = f + \Delta f$. Strain ε can be evaluated on the gauge length f, if the amount of change in f, Δf, can be determined.

For two parallel gratings it is valid that [eqn. (2.12)]

$$f = p\frac{(1 + \lambda)}{\lambda}.\tag{10.2}$$

When one of the gratings is deformed so that the interfringe spacing changes from f to f' it is valid that

$$\frac{f'}{p(1+\varepsilon)} - \frac{f'}{p(1+\lambda)} = 1,$$

from which it is deduced that

$$f' = p\frac{(1+\varepsilon)(1+\lambda)}{(\lambda-\varepsilon)} \quad \text{and} \quad \Delta f = p\varepsilon\frac{1+\lambda^2}{\lambda(\lambda-\varepsilon)}.$$

Term $p\varepsilon\lambda^2$ is an infinitesimal of higher order than that of term $p\varepsilon$ and may be neglected. Thus,

$$\Delta f = \frac{p\varepsilon}{\lambda(\lambda-\varepsilon)} \quad \text{and} \quad \frac{\Delta f}{f} = \frac{p\varepsilon}{\lambda(\lambda-\varepsilon)} \cdot \frac{\lambda}{p(1+\lambda)}, \tag{10.3}$$

which, by neglecting terms of higher order, reduces to

$$\frac{\Delta f}{f} = \frac{\varepsilon}{(\lambda-\varepsilon)}. \tag{10.4}$$

Introducing the value for λ given by eqn. (10.2) in eqn. (10.4) it is deduced that

$$\Delta f = \frac{\varepsilon f^2}{p - f\varepsilon}. \tag{10.5}$$

In the case where gauge length f is taken equal to 1 in. relation (10.5) becomes

$$\Delta f = \frac{\varepsilon}{(p-\varepsilon)}. \tag{10.6}$$

By using relation (10.6) it is possible to trace a family of curves of variations Δf in terms of the externally applied deformation ε for various parameters of pitch p of the grating in the case where the gauge length is kept constant and equal to 1 in. These parametric curves are shown in Fig. 10.1. Similar curves have been given by Dantu.[1] These curves present a discontinuity for values

[1] Dantu, P., Laboratoire Central des Ponts et Chaussées, Publ. 57-6 (1957).

of strains ε equal to pitch p of the gratings. In this case the moiré pattern becomes a uniformly bright field and the interfringe spacing becomes infinite. The difference, therefore, between the constant interfringe spacing of 1 in. and the interfringe spacing for this case is infinite. For values of strains ε larger than pitch p relation (10.6) must be modified by subtracting from ε the integral multiples of p.

Figure 10.2 shows a parametric family of curves of the minimum

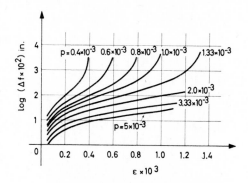

Fig. 10.1. Parametric family of curves of variation of the interfringe spacing Δf in terms of the externally applied strain ε for various values of pitch when the gauge length f is kept equal to 1 in.

detectable external strains ε in terms of variable gauge lengths for various parameters of pitch p of the gratings yielding a minimum interfringe increment $\Delta f = 0.05$ in.

It can be seen from these curves that for a minimum detectable variation of the interfringe spacing equal to 0.05 in., a strain of 10^{-4} can be estimated on a gauge length equal to 0.53 in. with gratings having a pitch $p = 0.6 \times 10^{-3}$ in. For the same strain $\varepsilon = 10^{-4}$ a gauge length $f = 1.2$ in. is necessitated if the pitch of the gratings is increased to $p = 3 \times 10^{-3}$ in.

Similarly, Fig. 10.2 shows a diagram of the limiting value of strain ε which can be detected on a gauge length $f = 1$ in. for various values of pitch p, by considering that a variation of inter-fringe spacing equal to one five-hundredth of the gauge length can be detected.

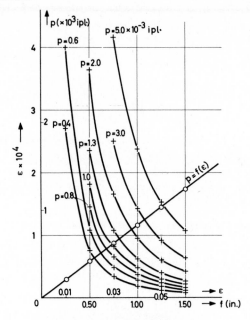

Fig. 10.2. Parametric family of curves of the minimum detectable strains ε in terms of variable gauge lengths for various parameters of pitch p yielding a minimum interfringe increment $\Delta f = 0.05$ in. Also shows the limiting values of strain ε which can be detected on a gauge length $f = 1$ in. for various values of pitch p.

This precision of measurements can be attained by direct observation of the moiré patterns formed by two interfering gratings. This precision can be further increased and fractions of fringes can be detected by a photometric method by using the continuous light-intensity displacement law. This increase of precision was discussed in §§ 2.8 and 2.9. The precision of measurements may be further increased by taking care to sharpen the moiré fringes.

The transmittances of the reference and the specimen gratings are defined as

$$T_r = \frac{p_B}{p - p_B},\qquad(10.7)$$

$$T_s = \frac{p'_B}{p' - p'_B},\qquad(10.8)$$

where p and p' are the pitches of the reference and the specimen gratings respectively and p_B and p'_B are the widths of the opaque strips of these gratings.

Similarly, the transmittance of the moiré fringes is given by

$$T_M = \frac{f_B}{f - f_B},\qquad(10.9)$$

where f_B is the part of the interfringe spacing covered by the black fringe.

From simple geometric considerations of the triangles formed by the gratings considered as consisting of successive opaque and transparent stripes and the moiré fringes it can be shown that

$$\frac{f_B}{f} = \left(\frac{p_B}{p} + \frac{p'_B}{p'} - 1\right).\qquad(10.10)$$

In the limiting case where the pitches p and p' of the two gratings are equal and their transmittances are equal, relation (10.10) becomes

$$\frac{f_B}{f} = \frac{2p_B - p}{p}\qquad(10.11)$$

and

$$\frac{f_B}{-f_B} = \frac{(p - 2p_B)}{2(p_B - p)}.\qquad(10.12)$$

Relation (10.12) indicates that the moiré fringe width f_B becomes zero when $p_B = p/2$, that is when the transmittance of the gratings is 50 per cent. (See also § 2.9.) The moiré fringe width becomes positive for $p_B > (p - p_B)$ and negative for $p_B < (p - p_B)$. Similar results have been presented by Zandman, Holister and Brcic.[2]

It is worth while noting that the moiré fringe width f_B, which resulted from the geometric construction, is defined by the region where light extinction is complete along the fringe length. The apparent fringe width, as seen by eye, is larger than that calculated from eqn. (10.10).

The best results for fringe sharpening, without losing in variation of light intensity from light to dark fringes, can be obtained with a transmittance equal to 50 per cent, that is when the bar and the corresponding slit widths of the gratings are equal.

Since the variation of light intensity from light to dark fringes is not so crucial in defining a moiré pattern it is possible, in some cases, to increase the sharpening of moiré fringes by using complementary gratings. In this case it is valid that

$$\frac{p_B}{(p - p_B)} = \frac{(p' - p'_B)}{p'_B}, \tag{10.13}$$

which yields

$$\frac{p_B}{p} + \frac{p'_B}{p'} - 1 = 0. \tag{10.14}$$

Comparing eqns. (10.10) and (10.14) it results that $f_B = 0$, that is the geometric moiré fringe width for complementary gratings is equal to zero. The fringes formed are, therefore, ideally sharp and narrower than in the case of moiré fringes formed by gratings having a transmittance 50 per cent. This effective sharpening of moiré fringes produced by using complementary gratings may be used as a further means for determining smaller fractions of moiré fringes. As an example the moiré sharpening was used by Holister, Jones and Luxmoore to obtain the possibility of locating

[2] Zandman, F., Holister, G. S., and Brcic, V., *J. Strain Anal.* **1** (1) 1 (1965

the moiré fringes formed at the ground-glass screen of their extensometer[3] to within $\pm 0{\cdot}01$ of the interfringe spacing.

10.2. Influence of grating defects on moiré fringes

The theory of interpretation of moiré fringes developed in the foregoing chapters may be normally applied to cases:

(a) When the deformed specimen grating remains in its initial plane after deformation.

(b) The planes of the specimen grating and the reference grating are coincident or parallel between them with an infinitesimal distance separating them.

(c) There is no relative angular displacement between the plane of the reference grating and the deformed specimen grating.

In the following the influence of slight deviations from the above-mentioned ideal conditions will be studied as well as the influence of periodic irregularities in the rulings of the gratings on the precision of the results obtained by the moiré patterns.

10.2.1. *Influence of a lateral deformation of the gratings*

Let O be the origin of the Cartesian coordinate system, Oxy be the plane of the specimen grating and Oz the axis normal to this plane at a generic point O of the grating (Fig. 10.3). The Oz axis coincides with the optical axis of the photographic apparatus used for recording the moiré pattern and C is the optical centre of the camera.

Let a generic point M of the specimen grating be displaced during deformation of SG by the components u, v and w along the axes Ox, Oy and Oz respectively. It is of interest to study the influence of the normal component w of displacement on the image of the grating as it was detected by the photographic apparatus. It is convenient to accept that displacement w is infinitesimal because it corresponds to reality and simplifies considerably the calculations. In any case this displacement cannot be larger than

[3] Holister, G. S., Jones, W. E. M., and Luxmoore, A. R., *Strain*, **2** (4) 27 (1966).

the depth of field of the objective, so that the image of the grating on the photographic plate may be considered as a central projection of the deformed surface of SG.

If MM' is the normal displacement w of point M, this displacement is equivalent from the point of view of photographic observation to a displacement MM'' in the Oxy plane. The displacement MM'' has as components along the Ox and Oy axes the quantities Δu and Δv.

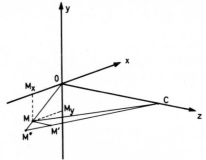

FIG. 10.3. Geometry of displacements of the projections of rulings of a grating due to a lateral deformation in the plane of the grating.

It can be readily found from simple geometric considerations that

$$\Delta u = w\frac{x}{D},$$
$$\Delta v = w\frac{y}{D}, \tag{10.15}$$

where D is the distance of the optical centre of the camera from origin O.

The relative corrections $\Delta\varepsilon_x$, $\Delta\varepsilon_y$ and $\Delta\gamma_{xy}$ to the components of strains ε_x, ε_y, γ_{xy} due to the lateral displacement w are given by

$$\Delta\varepsilon_x = \frac{\partial}{\partial x}\Delta u,$$

$$\Delta\varepsilon_y = \frac{\partial}{\partial y}\Delta v, \qquad\qquad (10.16)$$

$$\Delta\gamma_{xy} = \frac{\partial}{\partial y}\Delta u + \frac{\partial}{\partial x}\Delta v.$$

Relations (10.16) yield for $\Delta\varepsilon_x$, $\Delta\varepsilon_y$ and $\Delta\gamma_{xy}$ the quantities

$$\Delta\varepsilon_x = \frac{w}{D} + \frac{\partial w}{\partial x}\frac{x}{D},$$

$$\Delta\varepsilon_y = \frac{w}{D} + \frac{\partial w}{\partial y}\frac{y}{D}, \qquad\qquad (10.17)$$

$$\Delta\gamma_{xy} = \frac{\partial w}{\partial y}\frac{x}{D} + \frac{\partial w}{\partial x}\frac{y}{D}.$$

By denoting α and β the components of angular displacement of the element of surface in the neighbourhood of point M it is valid that

$$\alpha = \frac{\partial w}{\partial x} \quad \text{and} \quad \beta = \frac{\partial w}{\partial y}. \qquad (10.18)$$

Then relations (10.17) become

$$\Delta\varepsilon_x = \frac{1}{D}(w + \alpha x),$$

$$\Delta\varepsilon_y = \frac{1}{D}(w + \beta y), \qquad\qquad (10.19)$$

$$\Delta\gamma_{xy} = \frac{1}{D}(\beta x + \alpha y).$$

In the case of a two-dimensional specimen deformed in its plane by externally applied loads component w of displacement is originated from thickness variations of the specimen due to Poisson's effect. This thickness variation $2w$ in a specimen with an initial thickness equal to h is given by

$$\frac{2w}{h} = -v(\varepsilon_x + \varepsilon_y),$$

where v is Poisson's ratio of the material of the specimen. Then

$$w = -\frac{vh}{2}(\varepsilon_x + \varepsilon_y).$$

This deformation is always very small and its influence on the moiré pattern insignificant except in the neighbourhood of the specimen, where external concentrated loads are applied. Always large plastic deformations appear at these areas and their effect is perceptible, since the components of angular displacement α and β of the tangent plane to the deformed lateral surface of the specimen take large values. In order to reduce the correction calculations it is possible to place the point of application of the concentrated load on the optical axis of the equipment. In this case the contribution of α and β is annulled and term w/D only remains, which is given by

$$\frac{w}{D} = -v\frac{h}{2D}(\varepsilon_x + \varepsilon_y).$$

The relative value of the correction is equal to

$$\frac{\Delta\varepsilon_x}{(\varepsilon_x + \varepsilon_y)} = \frac{\Delta\varepsilon_y}{(\varepsilon_x + \varepsilon_y)} = -v\frac{h}{2D}. \qquad (10.20)$$

This relative correction is rather reduced, being of the order of 1 per cent and may be neglected. Moreover, in the case of several simultaneously applied concentrated loads to a specimen it is possible to bring successively each such point to the optical axis of the apparatus and take one photograph of the moiré pattern which is very close to the true distribution of displacements at the vicinity of each point of load application.

In the case of a three-dimensional specimen, where the applied loads are not symmetric to the grating plane, the normal components of deformation of the body may be perceptible and the errors derived from neglecting the terms with the components of angular displacement α and β may be large. An easy procedure of circumventing the difficulty and, in many cases, the impossibility of determining these terms all over the field, is to increase distance OC between the specimen and the optical centre of the camera and divide the meridian section of the specimen into restricted areas which will be photographed separately by placing the optical axis of the photographic apparatus at the centre of each area.

10.2.2. *Influence of a non-coincidence of the planes of reference and specimen gratings*

Consider the case where both the reference and the specimen gratings remain plane but the specimen grating is placed at some distance apart from the reference grating and is inclined. Let $\delta = OO'$ be the distance between the two gratings along the optical axis and let the reference grating occupy the Oxy plane. Again, let α and β be the components along the Ox and Oy axes of the angular displacement respectively of the plane of specimen grating with respect to the reference grating. Quantities δ, α and β are considered as infinitesimal.

The total lateral displacement between the planes of the two gratings may be expressed as

$$w = \delta + \alpha x + \beta y, \tag{10.21}$$

from which it is deduced that

$$\left. \begin{aligned} \Delta u &= \frac{x}{D}(\delta + \alpha x + \beta y), \\[2mm] \Delta v &= \frac{y}{D}(\delta + \alpha x + \beta y), \end{aligned} \right\} \tag{10.22}$$

and

$$\Delta\varepsilon_x = \frac{1}{D}(\delta + 2\alpha x + \beta y),$$

$$\Delta\varepsilon_y = \frac{1}{D}(\delta + \alpha x + 2\beta y),$$ (10.23)

$$\Delta\gamma_{xy} = \frac{1}{D}(\beta x + \alpha y).$$

Consider now the lateral displacement of the specimen grating relative to the reference grating with zero inclination. For parallel grating planes it is valid that

$$\Delta\varepsilon_x = \Delta\varepsilon_y = \frac{\delta}{D} \quad \text{and} \quad \Delta\gamma_{xy} = 0. \qquad (10.24)$$

Relations (10.24) show that it suffices to add a constant correction to strain terms ε_x and ε_y in order to eliminate the error due to a relative distance between the gratings. The error corresponding to the uniform strain field defined by eqns. (10.24) is equivalent to the initial moiré pattern formed by the line gratings superposed face to face ($\delta = 0$) with parallel principal directions and with the reference grating having a linear disparity in pitch equal to

$$\lambda = \frac{\delta}{D}. \qquad (10.25)$$

In this case the initial moiré pattern consists of equispaced parallel moiré fringes. Moreover, the moiré pattern is formed outside the planes of the gratings, as has been discussed in § 5.2, and at a distance l from the mid-plane between the planes of the two gratings, given by

$$l = \frac{\delta}{2}\frac{\lambda + 2}{\lambda}. \qquad (10.26)$$

The interfringe spacing of this initial moiré pattern is given by [see also eqn. (5.10)]

$$f = \frac{p(1+\lambda)}{\lambda}. \tag{10.27}$$

In the case where an inclination between the two gratings exists, it may be accepted without losing generality that the axis of inclination coincides with the Ox axis. In this case it is valid that $\alpha = 0$ and $\delta = 0$. Therefore

$$\left.\begin{aligned}
\Delta\varepsilon_x &= \beta\frac{y}{D}, \\[2mm]
\Delta\varepsilon_y &= 2\beta\frac{y}{D}, \\[2mm]
\Delta\gamma_{xy} &= \beta\frac{x}{D}.
\end{aligned}\right\} \tag{10.28}$$

In this case the theory developed in § 5.5 for inclined gratings having different pitches ($\lambda \neq 0$) is valid. The moiré pattern is formed in a plane passing through the intersection of the planes of the two gratings (Ox axis) and forming an angle φ with the specimen grating plane given by relation (5.49).

Since in both cases the effective pitch of the reference grating is larger than the pitch of the specimen grating, both moiré patterns for the cases of parallel planes and inclined planes of the gratings are virtual and they are formed behind the specimen grating.

In addition, the theory introduced in this subsection is rigorously valid only in the case where distance D is infinite. Its satisfactory validity remains for small relative angles of view of the specimen, that is in cases where the overall linear dimensions of the specimen are small compared to the optical distance between the camera lens and the specimen. For large angles of view of the specimen, distance $OC = D$ cannot be taken constant all over the field and the validity of the above given formulae is reduced.

A remedy in cases where corrections due to errors of orientation of the gratings cannot be evaluated, is to define the components of strains at free boundaries and corners of the specimen. Along the free boundaries it is valid that

$$\gamma_{nt} = 0 \quad \text{and} \quad \varepsilon_n + v\varepsilon_t = 0, \tag{10.29}$$

where subscript n defines a direction normal to the boundary, while subscript t defines a direction tangent to the boundary. By calculating the components of strains at free boundaries, it is possible to estimate the necessary corrections all over the field.

10.2.3. *Influence of an angular displacement of the reference grating*

The influence of an angular displacement α of the reference grating relative to the specimen grating was discussed in § 2.2. It is valid that

$$\left. \begin{array}{l} \Delta u = \alpha y, \\ \Delta v = -\alpha y, \end{array} \right\} \tag{10.30}$$

and, therefore,

$$\Delta\varepsilon_x = \Delta\varepsilon_y = 0.$$

If the angular misalignments in two orthogonal directions along which the reference grating must be successively superposed to yield the terms of the shear component of strain are equal and of the same sign, then the experimental error in defining the shear component γ_{xy} is zero and it is valid that

$$\Delta\gamma_{xy} = 0.$$

In this case, while the moiré pattern is considerably modified due to the angular displacement, the interpretation of the moiré pattern by graphical differentiation yields the same components of strain without the necessity of making any correction to the results obtained by the moiré patterns.

10.2.4. *Influence of periodic irregularities of gratings*

Consider a grating having a periodically variable pitch. Assume that the pitch is varying periodically over k intervals and in the range $(p \pm q)$. Let also the periodic variation of the pitch be sinusoidal, expressed by

$$y = mp + q \sin \frac{2\pi m}{k}. \qquad (10.31)$$

When the specimen grating is deformed this periodic variation becomes

$$y' = m'p + q \sin \frac{2\pi m'}{k} + v, \qquad (10.32)$$

where m and m' are integers $(m \neq m')$.

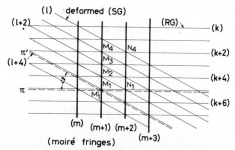

FIG. 10.4. Formation of moiré fringes by line gratings of equal pitch p angularly displaced through an angle ϑ where both gratings present a periodic error in pitch.

If two line gratings of equal pitch p are angularly displaced by an angle ϑ, and if there are no irregularities in the pitch of the gratings, the moiré fringes are formed along curves $M_1 M_2 \ldots M_n$, $N_1 N_2 N_3 \ldots N_n$, etc. Points $M_1, M_2 \ldots M_n$ undergo a slight displacement due to the sinusoidal variation in pitch (Fig. 10.4).

Consider a generic point M_1 at the intersection of the nth ruling of the reference grating and the n'-th ruling of the specimen grating both being considered perfect. Rulings n and n' undergo normal displacements because of the irregularities in pitch equal to

$$q \sin \frac{2\pi n}{k} \quad \text{and} \quad q \sin \frac{2\pi n'}{k}$$

respectively. Vector $M_1 M_1'$ connecting the corresponding points of intersection of rulings n and n' in the perfect and the periodically varying gratings is given by

$$\overline{(M_1 M_1')} = \frac{q}{\vartheta}\left(\sin\frac{2\pi n}{k} - \sin\frac{2\pi n'}{k}\right)\bar{i} + q\sin\frac{2\pi n}{k}\bar{j}, \quad (10.33)$$

where $\bar{i}$ and $\bar{j}$ are the unit vectors along the principal and secondary directions of the reference grating and ϑ the acute angle formed by the principal directions of the gratings at point M_1.

While the vertical component $\bar{j}$ always remains weak and contributes an insignificant amount of shifting in moiré fringes, component $\bar{i}$ parallel to the rulings of the reference grating may become important because of the contribution of factor $1/\vartheta$. The coefficient of this term may be expanded,

$$A = \frac{2q}{\vartheta}\sin\frac{2\pi}{k}(n-n')\cos\frac{2\pi}{k}(n+n'). \quad (10.34)$$

In the case where the subtractive moiré pattern is effective the quantity $(n-n')$ remains constant along any moiré fringe. If the additive moiré pattern is effective then the sum $(n+n')$ remains constant. In any case, either of the trigonometric coefficients in relation (10.34) remains constant along the fringes and may be replaced by

$$(n \pm n') = c,$$

when relation (10.34) becomes either

$$A_\alpha = \frac{2q}{\vartheta}\sin\frac{2\pi c}{k}\cos\frac{2\pi}{k}(c+2n'), \quad (10.35)$$

$$A_s = \frac{2q}{\vartheta}\cos\frac{2\pi c}{k}\sin\frac{2\pi}{k}(c-2n'). \quad (10.36)$$

In these relations the only variable quantities are angle ϑ and integer n'.

Relations (10.35) and (10.36) are periodic functions with a period equal to $k/2$ and an amplitude either

$$\frac{2q}{\vartheta}\sin\frac{2\pi c}{k} \quad \text{or} \quad \frac{2q'}{\vartheta}\cos\frac{2\pi c}{k},$$

which is a function of angle ϑ and constant c.

The troughs of these periodic functions correspond to

$$\frac{2\pi}{k}(c\pm 2n')m\pi, \tag{10.37}$$

where m is an integer.

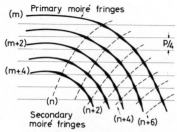

Fig. 10.5. Formation of secondary moiré phenomena due to periodic errors in the pitch of two superposed gratings.

It may be deduced from the above analysis that the troughs of the periodic functions A_a or A_s correspond to the intersections of moiré fringes with the rulings of the deformed grating spaced at a distance $kp/4$, i.e. at the quarters of the period of irregularities. These troughs, which correspond to maxima for the functions A_a and A_s, are manifested by periodic maxima in the width of the fringes and form a secondary moiré pattern aligned according to the diagonals of the primary moiré fringes (Fig. 10.5).

While along a certain primary moiré fringe the amplitude of troughs depends on angle ϑ, by leaping from one fringe to its neighbour the amplitude of troughs varies according to $\sin(2\pi c/k)$.

There are therefore primary moiré fringes without waving forms indexed by the even orders of the half wavelength and fringes with a maximum waving indexed by the odd orders of the half wavelength of the periodic function. Typical examples of the influence of the periodic irregularities of the gratings on the formation of moiré fringes are given in the moiré patterns shown in several figures given by Durelli.[4]

[4] Durelli, A. J., *Proc. Second Int. Cong. Exp. Mech.*, B. E. Rossi (Ed.), Soc. Exp. Stress Analysis, Westport, Conn., 1966, p. 1.

CHAPTER 11

Technological Information on Reproduction Techniques of Gratings

11.1. Production of large field gratings

A primary requisite for the application of moiré methods in the field of strain analysis is the possession of a good quality master grating ruled to the desired line frequency.

Photographic copies of grids utilized in the photoprinting technique are suitable for gratings of large pitch. The largest frequencies of gratings used by the photoengravers are approximately 200 lines per inch. These gratings are appropriate for the measurement of large plastic deformations in ductile materials.

It is possible to photographically reduce a coarse master grid and to obtain a good quality dense grating. Gratings of frequencies of the order of 500 lines per inch may be made by a single reduction. In this case a special photoengraver's process camera is needed for the accurate reproduction of gratings. The apparatus used in the laboratory for Testing Materials of the Athens Technical University was a camera supported on a main metallic bed of a length of 180 in. supported on four rubber absorbers thus insulating the apparatus from random vibrations of the floor (Fig. 11.1).

On the main bed of the apparatus a camera is located on a base disposing mechanisms for the parallel movement of the camera front and back. The parallel motion is effected by a double rack-and-pinion arrangement which guarantees the accuracy of the displacement of these elements. The lens housing is equipped with means for accurate vertical and horizontal motion of the lens for centring the image. The copyholder easel is placed at the other extremity of the camera bed and possesses means of parallel travelling as well as for angular adjustments. It is equipped with transparency holder as well as with a section copyholder when the object to be copied is opaque and flexible. The illumination of

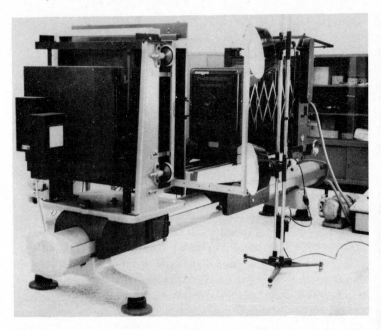

Fig. 11.1. Photographic view of a photoengraver's process camera.

opaque objects is achieved by the reflection of light produced by mercury vapour lamps fitted on travelling pedestal stands. In the case of transparencies a uniformly diffused xenon-light is used.

The lenses used in this camera are all of the apochromatic process type and cover a wide range of focal lengths. Process lenses are designed for magnifications of up to 1–3 and are corrected to a high degree within tolerances of the order of ± 0.01 per cent measured from the centre of the image to the edges.

The reduction possibilities of the camera are 1–5 while the magnification goes up to 2·2–1.

For a faithful reproduction of the master grating exact parallelism between the master grating plane, the lens principal plane and the image plane was required. This was achieved by removing the lens and the object and placing a flat mirror in the position of the object and another flat mirror with pinholes, arranged along two orthogonal straight lines centrally disposed, in the plane of formation of the image. The parallelism of these three planes can be achieved by the well-known method of multiple reflections.

The camera and the parallelism method of the object, lens and camera back were used either for reproduction of master gratings with some amount of reduction in pitch or for the image interference moiré method, where the image of the specimen grating on the ground-glass screen of the camera interfered with the reference grating.

Recently, ruling machines were introduced which can trace up to 270,000 grooves per inch. Normally traced gratings have from 7500 to 30,000 grooves per inch with an absolute uniformity of spacing. Periodically repeating errors exceeding $0·1$–$0·5$ μ in. are not tolerable in these master gratings. These gratings make ideal masters but their cost runs very high. The practice in this case is to produce replicas from the master by a precision moulding process and use the replicas as masters in each laboratory. However, the cost of the replicas is also high (a ruled area $2·5 \times 2·5$ in. with 91,500 grooves per inch sells for about \$1200). However, from the point of view of precision the replicas are superior to the masters.[1]

The ruling machine basically consists of a periodic reciprocating ruling head that holds a diamond tool above a carriage which supports the grating. A high precision screw indexes the carriage in relation to the tool. The machine disposes a barometric pressure compensation system maintaining a stable environment for long periods of time and phase correction and amplitude control systems compensating screw-pitch errors which exceed the errors

[1] Loewen, E. G., *Control Engng.* **15**, 95 (1963).

permitted on the grating. These ruling machines yield excellent master gratings with frequencies of several thousand lines per inch and in large overall dimensions (15–20 in. sides).

For the case of orthogonally crossed gratings a very adequate method for obtaining excellent master gratings is the electroforming process. The process is currently used for producing direct view storage tubes, image orthicon, vidicon and other special purpose vacuum tubes. The process consists of producing copper, silver, gold or any other metal screens by evaporating under vacuum the metal on to glass plates. For obtaining a thickness of the grating of several microns the glass plate is previously covered by a metallic layer. A tracing machine engraves the furrows and removes the metal in these areas. The glass is attacked by fluorhydric acid, which hollows the furrows. Before the evaporation the grating is submitted to various treatments which facilitate the separation of the screens. The metal is evaporated on the grating at high vacuum and fills the furrows of the metallic layer. The screen is afterwards separated from the grating by an appropriate rinsing and is submitted to an annealing process which results in an expansion and amalgamation of crystals.

High quality crossed screens may be produced by a photochemical procedure. A photographic negative of the grating is used as a base. A colloid layer is spread over the base of the film and insulation is achieved through the negative. The non-insulated areas are dissolved. The material of the base, which remains unmasked over the regions where the colloid layer is not exposed to light, was immersed in an electrolyting bath. The screen formed by electrolysis is separated from the base at the end of the procedure.

Both procedures yield excellent crossed gratings with a predetermined value in transmittance. Fifty per cent transmittances may be easily achieved. Each hole in the screen has perfectly square intersections and the slit and bar sizes may be controlled to tolerances of $\pm 2\ \mu$. Similarly, the tolerances at the right angles are very tight, yielding excellent orthogonally crossed gratings. It has been common practice in the laboratory for testing materials of the Athens Technical University to use such master gratings in cases where crossed gratings are needed in applications.

An interesting method was recently introduced by Fidler and Nurse[2] to produce large master gratings from long linear diffraction gratings (supplied by the National Engineering Laboratory, England) employed in machine tool control. These strips are 12 in. long and 1 in. wide and are produced in frequencies of 500, 1000 and 2000 lines per inch printed on photographic glass plates with their rulings parallel to their width.

These gratings were traversed across a photographic plate, while being illuminated by collimated light. This traverse combs out a similar grating of a width equal to the traverse. For the transversal movement of these gratings a special machine was designed, the essential part of which was a carriage supported on two air bearings which allowed the carriage to float freely and smoothly along a cylindrical bar which constitutes the carriage guide. The instrument was shielded from any extraneous light and allowed the light to fall only on that part of the photographic plate which lay underneath the machine control grating. A towing mechanism driven by an electric motor allowed a uniform traverse of the carriage.

The ruling of a large area grating was made by setting the master narrow strip grating in the carriage and adjusting it so that its rulings coincided with the direction of travel. A photographic plate was placed with its emulsion side upwards on the base of the machine and the carriage adjusted to leave a clearance between the emulsion and the master of the order of 0·005 in. One traverse of the carriage, whilst exposing the photographic plate to collimated light, produced a large line grating of the same pitch as the master. Gratings with frequencies of 500, 1000 and 2000 lines per inch were produced in this manner having average dimensions 10×10 in. The quality was checked by comparing the master strip with the ruled grating. The absence of any moiré pattern indicated the absence of pitch error or deviations from straight lines.

[2] Fidler, R., and Nurse, P., *Experimentelle Spannungsanalyse*, VDI-Berichte, No. 102, 1966, p. 59.

A method of producing orthogonally crossed gratings, besides the direct reproduction from master screens created by the electro-forming process, is either by superposing two line gratings with their axes at right angles or by exposing the same line master grating twice successively on the same photographic plate after a relative angular displacement of the master by 90°. For the production of a satisfactory crossed grating it is essential to orthogonally cross the line masters. In order to achieve a high accuracy in defining the right angle of rotation a jig was used consisting of the photographic plate holder, which was surmounted by a rotating master-grating holder.[3] The master holder could freely rotate on a circular groove of the plate holder and could be fixed in the plate holder in two mutually perpendicular positions.

For printing a crossed grating the photographic plate was positioned in the plate holder with its emulsion upwards and the master line grating was placed in the grooves of the master holder and fixed by eccentrical adjustments. The plate was exposed under collimated light and the master holder was then angularly displaced to its respective 90° position. A second exposure of the same plate was then made. The plate, when developed, yielded a dotted grating which was the negative of the crossed grating. The same jig was used for the image moiré method where the photographic negative of the specimen grating interfered with the reference grating. By placing the reference grating in such a rotatable jig, it was possible with a crossed specimen grating and a line reference grating to form successively the moiré patterns along the two orthogonal directions by suitably rotating the reference grating to the interference positions.

A similar method was recently introduced by Diruy[4] in order to increase the line frequency of masters. For this purpose he used master gratings of a transmittance equal to 0·25 and a frequency equal to n. A first exposure by contact of the master on an unexposed plate reproduced one part of the grating. An appropriate device was used to give a translation of the master grating

---

[3] A similar jig was described by Fidler, R., and Nurse, P., *op. cit.*, p. 367.
[4] Diruy, M., private communication (February 1966).

relatively to the photographic plate to a direction parallel to the principal direction of the master grating and equal to the half-distance between the axes of two consecutive rulings. A second exposure executed under identical conditions as the first one produced a second family of rulings. Development of the doubly exposed photographic plate yields a grating of a doubled frequency relatively to the frequency of the master grating. If the exposure time was suitably selected a transmittance of 50 per cent was attained in the resulting gratings. The same method may be used for tripling and quadrupling the line frequency of a master grating without reducing the overall dimensions of the plate.

For the production of equispaced circular gratings Diruy[5] developed the method described below. A narrow cylinder was rotated about its axis by a synchronous motor and a gear box at slow adjustable speed of rotation. The cylindrical box brought a slit along a radius of the base facing the still camera of the apparatus. The slit was covered by a sector of a line grating with the principal direction of the rulings coinciding with the central line of the slit. The grating was illuminated by a diffuser made by a fluorescent tube and ground glass placed between the light source and the grating. Slow rotation of the cylinder in a dark room was producing on the photographic plate of the camera a circular grating with a frequency matching the frequency of the line grating.

A similar method was introduced by Fidler and Nurse.[6] The technique consisted essentially of rotating a photographic plate under a narrow slit of a line grating subjected to collimated monochromatic light. The machine constructed for this purpose consisted of a heavy rotor to which the photographic plate was attached, a body housing the rotor, a carrier for the strip of the grating and a microscope.

Since very lengthy exposures were anticipated, great stability in the rotating mechanism was essential. Air bearings were used since

[5] Diruy, M., Laboratoire Central des Ponts et Chaussées, Intern. Circ. (1961).

[6] Fidler, R., and Nurse, P., *J. Strain Anal.* **1** (2) 160 (1966).

they present a complete radial stiffness and a capability to maintain a constant centre of rotation. In order to take into consideration the skidding effect and to produce a grating of constant and of a 50 per cent transmittance the widening of open slits due to skidding must be the same at all radii. This was achieved by modifying the shape of the slit which had the shape of a curvilinear sector. The circular gratings produced by this method were of good quality and constant transmittance.

11.2. Reproduction of gratings to specimens

Once the submaster gratings were reproduced from the master grating it was necessary to select a specimen grating of the frequency required by the particular experiment and reproduce it on the surface of the specimen. From the several techniques of reproduction of gratings on specimens available two well-known processes will be described.

The first method employs Kodak Kodalith stripping films. Reproductions of the master grating on the stripping film are executed by contact. If the master grating is a crossed grating the first reproduction shows a negative image of the master grating representing a dotted grating. A reproduction of the negative image on the same type of film yields a positive copy of the grating which can be used as a specimen grating. The film in both cases must be developed in a special type of developer and rinsed in water for a limited period of time to avoid premature stripping of the film.

The dry film was cemented on the prepared surface of the specimen with the emulsion side facing the specimen. As cement, either Eastman 910 adhesive may be used or an epoxy resin unplasticized or plasticized to match the properties of the specimen material in cases where epoxy resin specimens are tested. After completely curing the adhesive the acetate base of the film must be peeled off leaving only the thin emulsion layer adhering on to the specimen surface.

Gratings used in this way on plastics were found to withstand large strains up to 15 per cent before failure of the emulsion occurred and at temperatures of up to 100°C. The stripping film reproduction technique may be used on flat or curved surfaces like cylinders, etc., without needing dark-room facilities.

The second method employs the Kodak Photo Resist (KPR). This is a bitumen based lacquer which is cured when exposed in ultraviolet light. When dried it is waterproof, acid resistant and alkali resistant. The liquid KPR is a dual-purpose coating material. It serves as an ink-receptive material for preparing surface-type photolithographic printing plates and as an acid-resisting material for various metal-etching techniques. Both methods have been used in moiré techniques. The first is applicable to room temperature tests, with metallic or other type of materials, while the second is the only convenient method for reproducing gratings for high temperature tests.

The preparation of the surface of the specimen is simple. After conventional polishing, the surface must be scrubbed with a pumice-compound or, alternatively, it must be thoroughly vapour degreased until it is capable of holding a film of water without breaking. The surface must then be chemically treated by an acid recommended for each metal for an electroplating or other metal-finishing operation. For special instructions see the pertinent reference pamphlets.[7] The chemically treated surface must be washed under running water and dried before spraying the sensitizer.

The coating or sensitizing of the surface of the specimen is executed by conventional techniques. The easiest way to apply a thin coating on the polished degreased and dried surface of the specimen is to spray the solution with a paint sprayer or to dip the component in a plate-coating machine, thus producing a coating of a thickness of the order of 0·001 in. as compared to the thickness of stripping layers which are about 0·005 in. As the coating is sensitive to ultraviolet light, the grating is therefore contact printed on the specimen by exposing the coating to a

7 Kodak Photosensitive Resists for Industry, Ind. Data Book P-7 (1962).

white-flame arc lamp. The exposure times of KPR coatings are much shorter than those of conventional coating techniques and depend on the light intensity, the coating thickness and the reflectance of the underlying material. After coating and exposure, the unexposed areas of the plate image are washed away with organic solvents contained in the special KPR developer. Thus, the grating is reproduced on the surface of the specimen as a cured and resistant lacquer. For this type of reproduction a negative crossed grating must be used, which consists of series of square dots. In this way the printed grating is a positive grating consisting of crossed lines. The grating may be further dyed by special dyes (blue or black) to increase further the contrast between the bars and slits of the gratings.

For temperatures higher than that the lacquer can withstand, an etching process is indispensable which etches the grating into the surface under test. While gratings reproduced on the specimen by photolithographic printing can withstand temperatures of up to 200°C the gratings reproduced by the etching technique can withstand temperatures of up to the corresponding melting points of the materials.

The KPR method was found to be superior in definition and in the ability of withstanding abrasion than conventional methods using photosensitive materials based on the photosensitivity of potassium dichromate in casein, albumin, shellac or other suitable materials. KPR is easy to apply and does not necessitate sophisticated dark-room facilities. The only disadvantage of KPR is its charring above 200°C and its oxidizing above 400°C. For high temperature tests the etching technique is essential. A supplementary treatment for the specimen, enhancing the quality of moiré patterns, may consist of an electric deposition of a different metal than the material of the specimen in the open slits of the grating. For steel specimens the surface may first be nickel-plated and a grating reproduced by the KPR method. This is succeeded by electric deposition of copper at the unexposed areas where the lacquer was washed away. Upon complete removal of the KPR lacquer and oxidation of the copper to darken the bars of the grating a satisfactory contrast is achieved yielding excellent moiré fringes at high temperatures.

11.3. Photographic equipment and materials

It was previously described that there are two distinct methods for obtaining the moiré patterns, i.e. the direct superposition and the image moiré methods. For the direct superposition method a good-quality photographic apparatus is necessary for recording directly the moiré patterns and there is no special need for elaborate photographic equipment. On the contrary, for the image moiré method it is necessary to faithfully reproduce a copy of the specimen grating on to the ground-glass screen of the camera and this obligation implies the use of high-resolution optical equipment free of any kind of aberrations and distortions of the image, which introduce spurious deviations in the moiré pattern formed by the image of the specimen grating and the reference grating. Experiments have shown that lens systems designed for small magnifications are adequate. Such systems are anastigmat lenses used in cases of enlarging processes or process apochromatic lenses. Process lenses are designed for reproductions of the order of unity and they are corrected to a high degree. Normal tolerances for such lenses may be of the order of ± 0.02 per cent from centre to the boundary of the image. A series of process apochromatic types of lenses with a wide range of focal lengths is required.[8]

The lenses must be attached to rigid cameras which possess an accurately adjustable distance between their lens front and their screen back. These planes must be also adjustable to angular displacements about two orthogonally crossed axes in order to secure an exact parallelism between the planes of object, the lens front and the camera back. Such photographic equipment is a photo-engraver's process camera which is shown in Fig. 11.1 (p. 380).

Photographic plates or films are normally used for recording moiré patterns from which the strain components in a deformation

[8] Kingslake, R., *Lenses in Photography*, 2nd edn., A. S. Barnes & Co. (Eds.), 1963.

field can be deduced from measurements of the interfringe spacing executed either by a travelling microscope or photomechanically by using a photocell.

The variation of the light intensity of the specimen grating, on which the contrast of the moiré pattern depends, relies on the reflectance or transmittance of the underlying material and the uniformity of the illumination. Since in most cases of materials studied these conditions are not satisfactory, the resulting moiré patterns present a poor contrast and the opacity over the moiré negatives is unevenly distributed. Therefore photographic materials of high contrast and sensitivity characteristics are needed in moiré work.

Orthochromatic films on estar base or orthochromatic plates of the lithographic series (e.g. Kodalith series) have an extremely high-contrast orthochromatic emulsion of moderate sensitivity, as well as an extremely high resolving power and extremely fine granularity. These films and plates are especially suitable for moiré pattern recording, as well as for reproduction of gratings. While the plate form is best suited to applications where dimensional stability and accurate positioning are extremely critical, the polyester film-base materials are adequate for all current photographic applications on moiré work since their polyester base exhibits very high tensile strength and resistance to tear and provides therefore the best dimensional stability available in a flexible support. For more details the reader is referred to the relevant bibliography.[9]

[9] See, for example, Eastman Kodak Co., *Kodak Plates and Films for Science and Industry*, Kodak Publ. No. P-9 (1962); *Films and Plates for the Graphic Arts*, Kodak Publ. No. Q-2 (1961).

CHAPTER 12

Evaluation of Moiré Methods

IN THE previous chapters the theoretical and technological aspects of the moiré methods were developed yielding an overall picture of the interpretation of moiré patterns as they are applied to strain analysis. Here an outline of the advantages of these techniques over those of conventional techniques in strain analysis will be presented.

One of the main advantages of the moiré methods is that they are able to determine directly strain distributions from purely geometric relationships derived from the interference of the specimen and reference gratings without depending on measurements of intermediate physical properties such as changes in resistance, inductance, capacitance as with electric strain gauges or birefringence in classical photoelasticity and birefringent coating methods. Moreover, the moiré methods yield the strain distributions in actual components of structures or models made of the same material as the prototype and hence they eliminate similarity conditions imposed by the use of models of materials different than those of prototype constructions.

Besides the geometrical and load distribution similarities between model and prototype, which are required for the correct transfer of the data from the model to structure, it is necessary to consider the mechanical characteristic properties of the materials of the model and prototype and their changes under different loading conditions.

In problems of elasticity, both Young's modulus and Poisson's ratio may be accepted as constant. If the ratios of constants E_m

and v_m of the model to E_p and v_p of the prototype remain constant over the range of loading in model and prototype, they do not influence the detailed picture of strain and stress distributions between model and prototype. It suffices, therefore, to determine the dimensionless strain and stress concentration factors in the model and transpose them without any alteration to the corresponding strain and stress distributions in the prototype.

The mechanical characteristic properties of the materials change drastically in space and time in plasticity, creep, relaxation, thermal, dynamic and in all other transient phenomena. In these cases all the similarity conditions cannot be in general simultaneously satisfied between model and prototype and therefore the methods employing models fail to yield the real picture of the strain and stress distribution in the structure and only approximate relationships can be introduced where the characteristic mechanical variables are considered as constant over a certain area in space and length of time.

These complex types of problems are of practical importance and for these the convenient moiré method, appropriately chosen, yields a direct answer of the strain and stress distribution in the prototype by considering the corresponding variation of the mechanical properties of the material of the structure.

In addition, the variations in moiré fringe patterns due to externally applied loads are directly related to the individual normal and shear strains referred to the specimen and reference grating principal directions, while in other strain analysis methods, such as photoelasticity, the components of principal stress or strain difference are determined and, therefore, complementary information is required for separating the individual stresses or strains.

The techniques used for the reproduction of gratings on to the component's surface are such that they do not damage the surface finish. The only exception being the case of surface etching which is used in cases of strain distributions in structure members subjected to steady or transient thermal fields. In these cases the etching depth can be controlled and therefore reduced to insignificant amounts in order that its influence may be assumed as negligible. In particular, if care is taken to finish the specimen surface satisfactorily flat and polished so that accidental scratches and pits are considerably reduced, and to illuminate the surface by a strong and collimated light beam, then shadows of the etching peaks on the troughs by the oblique illumination create a satisfactory contrast of the etched grating.

The etching technique is the only one suitable for the study of strain distribution in members subjected to a thermal field. This technique, combined with the image moiré method, forms a very suitable method for the study of the behaviour of materials at high and very high temperatures. In these domains any other conventional technique measuring strain fails to yield satisfactory results. Its incontestable merits derive from the fact that changes in the strain distribution are determined solely from the deformation of etched grating lines. Since no extraneous and alien bodies (such as an extensometer or electric strain gauge, etc.) come into contact with the heated specimen, there is no spurious influence on the heat distribution in the thermal field. In this manner not only steady-state thermal problems can be investigated but also transient state problems may be accurately solved. Therefore, the image moiré method remains the sole promising technique for an accurate evaluation of strain and stress distributions at high temperatures.

Since the strain distributions are determined from the photographic images of the rulings measurements can be obtained up to the red-glow or melt temperature of the material. This compares favourably with the range of accurate application of other conventional methods of strain measurements at high temperatures (such as strain gauges suitable for high temperatures) which, in general, are restricted within limits.

The specimen grating reproduction techniques, besides the etching technique, do not reinforce the structure element to be investigated as birefringent coatings, strain gauges, photoelastic gauges, etc. The photographic layer, on which the grating is reproduced, does not influence the strain and stress distribution in the field and in particular at free edges or strain raisers where the influence of reinforcement is most critical. Hence the values of strain yielded by moiré methods may be considered as approaching reality more than any other experimental method.

In general, conventional experimental methods either cause some damage on the surface finish (as are punch-marks, scratches and lateral restrain with mechanical and optical gauges which, being strain raisers, may well nucleate yielding or fracture) or measure deformations related to the strain distribution of the surface, which are influenced by the characteristics of the measuring device (as in the case of birefringent coatings where the coating thickness, edge effects and the effect of the dissimilar variation of the contraction ratio of the coating and the specimen must be taken into consideration for a more accurate evaluation of the results).

It is worth while stressing once more the importance of the image moiré method for measurements of displacement field in transient problems. In this case the specimen remains undisturbed from any contact with an alien body which may significantly influence the strain distribution in the specimen. Problems of creep, relaxation and recovery of viscoelastic materials may be studied in this way and the influence of external strain raisers may be readily detected during the evolution of the viscoelastic phenomenon and, if they cannot be completely eliminated, they can be taken into consideration during the evaluation of the strain distribution in the member.

The application of the image method to the study of transient phenomena, as it was exemplified in §§ 4.1, 4.2 and 7.3, where either the strain distribution in the Lüders instability region of low carbon steel was discussed or the geometrical instability of tension specimen due to necking, or the shock-wave propagation in three-dimensional transparent bodies was studied, yields a complete and accurate picture of the displacement field varying with time. No other experimental method could so accurately record the transient phenomena in these problems with such ease and minimum use of equipment. The extension of this method to shock-wave propagation phenomena in metals and other opaque substances in the study of the mechanical behaviour of such materials to high and very high pressures is a very promising field.

The analysis of the displacement field in a three-dimensional transparent model is another application where the image moiré technique presents an incontestable superiority over three-dimensional photoelasticity. In three-dimensional photoelasticity and in particular in the frozen stress technique (which is the most reliable three-dimensional photoelastic method) slices of the stress-frozen specimen are cut along different orientations and analysed in order to yield the exact picture of the stress distribution in the body. In the image moiré method when three specimen gratings oriented at 45° apart are used, it suffices to take three photographic records of the moiré patterns oriented at 45° in order to analyse the strain field in the body independently of the mechanical state of the material (elastic–plastic–viscoelastic).

In addition it is possible to incorporate in the specimen more than one specimen grating at different meridian sections of the body. By focusing the camera on to each one of these gratings it is possible to study the strain distribution at various sections of the transparent body. The image moiré method can be used with success in investigations of plastic strain distribution in interior sections of opaque bodies. This idea was used by Theocaris[1] in order to study the internal strain distribution in copper cubes subjected to pure and unlubricated compression. The copper cubes were sliced. On to the internal faces of the slices crossed gratings were reproduced by the etching technique; the slices were silver-soldered together and the cubes were subjected to compression. The results obtained after separating the slices of the deformed cubes gave the plastic strain distribution in the interior of the body with high accuracy.

Another advantage of the moiré techniques, which must not be underestimated, is the flexibility of the methods to determine changes in strain from small elastic deformations (linear and angular differential methods) to very large plastic deformations with the same simple means and with almost the same accuracy.

The fact that there is no fixed gauge length or location of gauge

[1] Theocaris, P. S., *Mechanics of the Solid State*, F. P. J. Rimrott (Ed.), 1968, pp. 240–258.

in moiré techniques, and that an evaluation of the components of strain may be carried out at any point in the ruled area and along any gauge length, is of great importance. The gauge length is a function of the initial line frequency of the gratings and the strain magnitude. A moiré pattern may therefore be considered as a cluster of gauges of a relatively small but varying gauge length placed the one close to the other, which yield simultaneously the values of strain components at any point in the area covered by the specimen grating. Hence the moiré methods are of great value in cases where the overall picture of the strain distribution over a large field are required and the nuclei of discontinuities in the strain distribution are not known in advance. Any discontinuity in the strain distribution is immediately apparent from disturbances occurring in the fringe patterns and can be precisely located and fully investigated. When large strain gradients are present, readings obtained by conventional experimental techniques yield an average value of the displacement over the gauge length of the instrument which can be very misleading. Such are the cases of appearance nucleation and spreading of plastic zone discontinuities in metals and the study of fracture phenomena in metals and other technical materials. These were treated in the examples studied of strain distributions in Lüders' discontinuities and in the nucleation of necking in a tensile specimen.

The moiré methods employing two gratings or similar devices, placed in remote parallel or inclined planes, present similar advantages to those of the image moiré method. While the principle of formation of a moiré pattern by remote parallel or inclined gratings is different to that of the image moiré method, both methods lead to the same results. As an example the two remote and parallel gratings method, termed in this text as "multisource method", was used with success for the study of crack accelerations and propagations in transparent tension members where one specimen grating was printed on the specimen and the reference grating was placed at some distance away from the specimen in order to avoid the effects induced by the contact of the reference grating with the specimen.

Besides this classical application, moiré methods based on the interference of remote gratings were used for the direct evaluation of the sum of principal stresses in two-dimensional transparent specimens. In this case the two gratings act as a lens and create an enlarged image of the gratings, which is distorted by the deformed specimen. The transmitted image moiré method, the multisource and the slit source and grating methods yield the isopachics of two-dimensional specimens loaded in their planes, while the reflected image moiré method, the multisource, and the slit source and grating methods give the patterns of partial slope contours in flexed plates. The methods proved to be very sensitive. They yield dense isopachic patterns in specimens made of brittle materials, as glass, where the lateral deformation of the specimen is insignificant.

Another promising application of the remote gratings for yielding directly the stress components in a deformed transparent specimen is to reproduce the two interfering gratings in both lateral surfaces of the specimen. A collimated light passing through the front surface of the specimen deviates according to the variation of the refractive index of the material, which changes proportionally with the stress field. The distorted image of the first grating interferes with the second grating and yields directly the components of stresses in the specimen.

The reflected image moiré method, as well as the multisource and the slit source and grating methods together with Ligtenberg's photoreflective and the Salet–Ikeda methods, are of real value in the study of flexed plates because they yield a detailed, accurate and clear picture of the partial slope distribution of flexed plates. As it is well known, it suffices from the partial slope distribution a single differentiation in order to obtain the Cartesian components of moments and strains in the plate and thus solve completely the problem of stress distribution in a flexed plate.

The reflected image moiré method was successfully extended to yield the nodal and antinodal configurations in vibrating plates as well as the deflection and phase distribution in the plate. The results obtained by this method applied to cases of vibrating plates, when compared to Chladni's sand figures, present a substantial progress in the study of vibration modes in plates. The extension of the application of the reflected image moiré method and the oblique shadow moiré method to the study of slope and deflection distributions in developable and ruled shells is of real value to a domain of applications where experimental methods yielding an overall picture of the deformation field of the shell are completely lacking and desperately needed. The results obtained in these applications do not only yield the slope or deflection distribution due to an externally applied load, but allow the measurement of the initial microscopic curvature, as well as its variation in the space of the shell.

Besides the application of the moiré patterns based on line gratings, methods were recently developed where circular, radial, zone or other types of gratings are used. Moiré patterns formed by the interference of radial or circular gratings present the advantage of yielding, besides the conventional information extracted from any moiré pattern formed by line gratings, a complete information concerning the definition of the vector of displacement of the one grating relatively to the other. This information is very important in the study of rigid-body movements in various parts of a construction.

The moiré gauges created by the interference of either two circular gratings (continuous rosette), or of a circular and a line grating, or the radial moiré gauges, when fully developed, may successfully compete with the electric or photoelastic strain gauges.

All moiré techniques are simple and easy to apply. They do not necessitate elaborate equipment. At the most, four photographs of moiré patterns after each loading step suffice for a complete and accurate evaluation of the strain distribution over the field independently of the state of stress and strain in which the specimen is subjected. While the recording of the strain field necessitates only fractions of a second, the calculation of the strain components may be executed either by skilled personnel far from the site of experiments helped by a desk calculator and disposing instruments for the accurate measurement of distances and angles, or automatically by feeding the data of the moiré pattern transformed in punched form to a combination of analog and digital computers.

The moiré methods lack in high sensitivity in cases where the classical equal-pitch moiré method is applied. But in cases where the sensitivity of measurements is very important the differential methods or the methods based either on the continuous light-intensity displacement law or on isodensitracing are very suitable, since their sensitivity is comparable to the sensitivities of the other conventional experimental methods.

References

A. Books

DALLY, J. W., and RILEY, W. F., *Experimental Stress Analysis*, McGraw-Hill, New York, Ch. 14.4, 1965.

DOVE, R. C., and ADAMS, P. H., *Experimental Stress Analysis and Motion Measurement*, C. E. Merrill Books Inc., pp. 21–31, 1964.

DUNCAN, J. P., Grid and moiré methods of stress analysis, *Stress Analysis*, Zienkiewicz, O. C., and Holister, G. S. (Eds.), J. Wiley, New York, Ch. 14, pp. 314–345, 1965.

DURELLI, A. J., PHILIPS, E. A., and TSAO, C. H., *Introduction to the Theoretical and Experimental Analysis of Stress and Strain*, McGraw-Hill, New York, pp. 306–308, 1958.

DURELLI, A. J., and RILEY, W. F., *Introduction to Photomechanics*, Prentice Hall, N.J., Ch. 8, 1965.

GUILD, J., *The Interference Systems of Crossed Diffraction Gratings*, Clarendon Press, Oxford, 1956.

GUILD, J., *Diffraction Gratings as Measuring Scales*, Oxford Univ. Press, London, 1960.

HOLISTER, G. S., *Experimental Stress Analysis, Principles and Methods*, Cambridge Univ. Press, pp. 249–286, 1967.

OSTER, G., *The Science of Moiré Patterns*, Edmund Scientific Co., Barrington, N.J., 1964.

RILEY, W. F., Moiré method of strain analysis, *Manual on Experimental Stress Analysis*, Typenny, W. H., and Kobayashi, A. S. (Eds.), Soc. Exp. Stress Anal. Publ., Westport, Conn., U.S.A., Ch. 6, 1967.

VALYUS, N. A., *Rastrobaia Optica*, Moscow–Leningrad, Ch. 15, pp. 366–379, 1949.

B. Papers

AITCHISON, T. W., BRUCE, J. W., and WINNING, D. C., Vibration amplitude meter using moiré fringe technique, *J. Sci. Instrum.*, Vol. 36, pp. 400–402 (1959).

ANDERSON, J. A., and PASTER, R. W., Ronchi's method of optical testing, Contribution from the Mount Wilson Observatory, Carnegie Inst. Washington, No. 386, pp. 175–181 (1929).

ANON., The Ferranti moiré fringe measuring system, Patent specification 760,321 (1954).

ANON., Strain measurement by moiré technique, *Engineering*, Vol. 180, p. 116 (1955).

ANON., Automatic machine tool control systems, *Aircr. Engng.*, Vol. 28, No. 329, pp. 244–246 (1956).

ANTHES, K., Versuchsmethode zur Ermittlung der Spannungsverteilung bei Torsion prismatischer Stabe, *Dinglers Polytech. J.*, pp. 342–345, 356–359, 388–392, 441–444, 455–459, 471–475 (1906).

BARAYA, G. L., PARKER, J., and FLOWETT, J. W., Mechanical and photographic processes for producing a grid of lines, *Int. J. Mech. Sci.*, Vol. 5, p. 365 (1963).

BARBER, B. L. A., and ATKINSON, M. P., Method of measuring displacement using optical gratings, *J. Sci. Instrum.*, Vol. 36, pp. 501–504 (1959).

BASAVA RAJU, B., A study of simply supported square plates with cut-outs by the moiré method and by finite differences, National Aeronautical Laboratory, Bangalore, Tech. Note No. TN–SA–S–64, pp. 1–13 (1964).

BASSET, G. A., MENTER, J. W., and PASHLEY, D. W., Moiré patterns on electron micrographs and their application to the study of dislocations in metals, *Proc. Roy. Soc. London*, A, Vol. 246, No. 1246, pp. 345–368 (1958).

BELL, J. F., Determination of dynamic plastic strain through the use of diffraction gratings, *J. Appl. Phys.*, Vol. 27, pp. 1109–1131 (1956).

BELL, J. F., 10,000 threads to the inch, *Am. Mach.*, pp. 112–113 (1956).

BELL, J. F., Normal incidence in the determination of large strain through the use of diffraction gratings, *Proc. Third U.S. Nat. Cong. Appl. Mech.*, pp. 489–493 (1958).

BELL, J. F., Diffraction grating strain gauge, *Proc. Soc. Exp. Stress Anal.*, Vol. 17, No. 2, pp. 51–64 (1960).

BERNARD, R., and PERNOUX, E., Franges d'interférances obtenues par la superposition de deux faisceaux électroniques cohérents, *Compt. Rend. Acad. Sci. Paris*, Vol. 236, pp. 187–189 (1953).

BLAKE, D. W., and SAYCE, L. A., Machine tool monitor using diffraction gratings, *Trans. Soc. Instrum. Tech.*, Vol. 10, No. 4, pp. 190–196 (1958).

BOUWKAMP, J. G., Orienterend experimentele Onderzoek d.m.v. de moiré methode waar de momentenverdeling in een paddestoelvloer bij verschillende belastigen, Afd. Wegen Waterbouwkunde, Delft, Technische Hogeschool, Report P–4 (1952).

BOUWKAMP, J. G., Analysis of two-dimensional stress problems by the moiré method, *Proc. First Int. Cong. Exp. Mech. New York*, pp. 195–218 (1963).

BOUWKAMP, J. G., The moiré method and the evaluation of principal-moment and stress directions, *Exp. Mech.*, Vol. 4, No. 5, pp. 121–128 (1964).

BRANDLEY, W. A., Laterally loaded thin flat plates, *Proc. ASCE, J. Engng. Mech. Div.*, Vol. 85, EM4, pp. 77–107 (1959).

BREWER, G. A., Measurement of strain in the plastic range, *Proc. Soc. Exp. Stress Anal.*, Vol. 1, No. 2, pp. 105–115 (1943).

BREWER, G. A., and GLASSCO, R. B., Determination of strain distribution by the photogrid process, *J. Aeronaut. Sci.*, Vol. 9, No. 1, pp. 1–7 (1941).

BREWER, G. A., and ROCKWELL, M. M., Measurement of the drawing properties of aluminium sheets, *Metal Prog.*, Vol. 41, No. 5, pp. 663–668 (1942).

BREWER, G. A., and ROCKWELL, M. M., Stress–strain relationships in the drawing of metals, *Metal Prog.*, Vol. 41, No. 6, pp. 806–810 (1942).

BROMLEY, R. H., Two-dimensional strain measurement by moiré, *Proc. Phys. Soc.*, B, Vol. 69, No. 3, pp. 373–381 (1956).

BURCH, J. M., The possibilities of moiré fringe interferometry, *Proc. Symp. Interfer.*, Paper 3–3, pp. 181–218 (1959).

BURCH, J. M., Photographic production of scales for moiré fringe applications, *Optics in Metrology* (Brussels Colloquium, May 6–9, 1958), Pergamon Press, pp. 361–368 (1960).

BURCH, J. M., The metrological applications of diffraction gratings, *Progress in Optics*, edited by E. Wolf (North Holland Publ. Co., Amsterdam), Vol. 2, pp. 75–108 (1963).

CASPER, W. L., Analysis of skew plates by the moiré method, M.S. thesis Research Report, Div. Struct. Engng. and Struct. Mech., Univ. of Calif., Berkeley (1960).

CHIANG, F. P., Method to increase the accuracy of the moiré method, *Proc. ASCE, J. Engng. Mech. Div.*, Vol. 91, EM1, pp. 137–149 (1965).

COOK, R. D., A method for measuring thickness changes in plane transparent models, *Exp. Mech.*, Vol. 5, No. 11, pp. 363–365 (1965).

CRISP, J. D. C., The measurement of plane strains by a photoscreen method, *Proc. Soc. Exp. Stress Anal.*, Vol. 15, No. 1, pp. 65–76 (1957).

DANIEL, I. M., Experimental methods for dynamic stress analysis in viscoelastic materials, *J. Appl. Mech.*, Vol. 32; *Trans. ASME*, Vol. 87, pp. 598–606 (1965).

DANIEL, I. M., Quasistatic properties of a photoviscoelastic material, *Exp. Mech.*, Vol. 5, No. 3, pp. 83–89 (1965).

DANTU, P., Determination expérimentale des déflexions dans une plaque plane, Mémoires des documents des Ingénieurs des Ponts et Chaussées, Paris, pp. 1–20 (1940).

DANTU, P., Méthode nouvelle de détermination des contraintes en élasticité plane, *Annls. Ponts Chauss.*, Vol. 110, No. 1, pp. 5–20 (1940); Vol. 122, No. 3, pp. 271–344 (1952), and No. 4, pp. 375–406 (1952).

DANTU, P., Recherches diverses d'extensométrie et de détermination des contraintes, *Analyse des Contraintes, Mém. GAMAC*, Vol. 2, No. 2, pp. 3–14 (1954).

DANTU, P., Utilisation des réseaux pour l'étude expérimentale des phénomènes élastiques et plastiques, *Compt. Rend. Acad. Sci. Paris*, Vol. 239, pp. 1769–1771 (1954).

DANTU, P., Utilisation des réseaux pour l'étude des déformations, Laboratoire Central des Ponts et Chaussées, Paris, Publ. No. 57–6 (1957), also published in *Ann. Inst. Tech. Bat. Trav. Publics*, Series: *Essais et Mesures*, Vol. 2, No. 121, pp. 78–98 (1958).

DANTU, P., Correction à apporter au principe de l'interprétation du moiré quand les déformations sont très importantes, Laboratoire Central des Ponts et Chaussées, Paris, Inter. Circ. (1960).

DANTU, P., Etude expérimentale de la déformation d'une sphère de plexiglas comprimé entre deux plaques parallèles rigides en dehors du domaine élastique, *Proc. Tenth Int. Cong. Appl. Mech., Brussels Univ.*, Paper 2–25 (1960).

DANTU, P., Application de la méthode des réseaux à l'étude de la déformation jusqu'à rupture d'éprouvettes en matières plastiques, *Proc. Int. Colloq. Rheol. Paris*, No. 98, pp. 1–12 (1960).

DANTU, P., Utilisation de la méthode du moiré pour l'étude des problèmes d'élasticité à trois dimensions, *Compt. Rend. Acad. Sci. Paris*, Vol. 258, pp. 4206–4207 (1964).

DANTU, P., Extension de la méthode du moiré à des problèmes thermiques. Etudes de déformations rémanentes d'un métal au voisinage d'un cordon de soudure, *Revue Fr. Méc.*, Vol. 2, Nos. 2–3, pp. 117–118 (1962); see also *Exp. Mech.*, Vol. 4, pp. 64–70 (1964).

DANTU, P., Moiré du deuxième ordre, méthode permettant d'obtenir directement les lignes d'égale dilatation linéique, *Analyse des Contraintes, Méms. GAMAC*, No. 17, pp. 55–62 (1966).

DATTA, S. K., On Brewster's bands, *Trans. Opt. Soc.*, Vol. 28, No. 4, pp. 213–217 (1926–27).

DAVIES, B. J., ROBBINS, R. C., WALLIS, C., and WILDE, R. W., A high resolution measuring system using coarse optical gratings, *Proc. Instn. Elect. Engrs.*, Vol. 107B, No. 36, pp. 624–633 (1960).

DECHAENE, R., Measurements of plastic strains in welded plates by the moiré method, Univ. of Ghent, I.I.W., Doc. 9, pp. 301–323 (1959).

DECHAENE, R., VAN HAUVAERT, J., and VYVEY, M., Meting von plastische vervormingen mit behulp van moiré-figuren, *Revue C, Génie Civ.*, Construction Bouwkunde Constructie, Vol. 1, No. 7, pp. 9–14 (1958).

DECHAENE, R., and VAN DE PUTTE, P., Measured plastic strains in notched plates, Univ. of Ghent, I.I.W., Doc. 10, pp. 323–362 (1959).

DE HAAS, H. M., and LOOF, H. W., An optical method to facilitate the interpretation of moiré pictures, *VDI-Berichte*, No. 102, pp. 65–70 (1966).

DE JOSSELIN DE JONG, G., Moiré patterns of the membrane analogy for groundwater movement applied to multiple fluid flow, *J. Geophys. Res.*, Vol. 66, No. 10, pp. 3625–3628 (1961).

DE JOSSELIN DE JONG, G., Refraction moiré analysis of curved surfaces, *Proc. Symp. Shell Res.*, Delft, pp. 302–308 (1961).

DEW, G. D., and SAYCE, L. A., On the production of diffraction gratings: I. The copying of plane gratings, *Proc. Roy. Soc. London*, A, Vol. 207, pp. 278–286 (1951).

DIRUY, M., L'analyse des contraintes par la méthode des réseaux optiques, *DOCAÉRO*, No. 55, March 1959.

DIRUY, M., Réalisation d'un extensomètre de haute sensibilité pour mesures sur grande base, *Révue Fr. Méc.*, Vol. 2, No. 4, pp. 101–112 (1962).

DOSE, A., and LANDWEHR, R., Zur bestimmung der Isopachen in der Spannungsoptik, *Naturwiss.*, Vol. 30, pp. 342–346 (1949).

DOSSCHE, M., Contribution à l'emploi des moirés en analyse expérimentale des contraintes, final dissertation, Univ. of Brussels (1959).

DOUGLAS, R. A., AKKOC, C., and PUGH, C. E., Strain-field investigations with plane diffraction gratings, *Exp. Mech.*, Vol. 5, No. 7, pp. 233–238 (1965).

DROUVEN, G., Measurement of the sum of the principal stresses in plane problems of elasticity by interference, Ph.D. dissertation, Washington Univ., St. Louis, Mo. (1952).

DUFFEY, H. J., and MESMER, G. K., Finite rotation and strain measurement using the moiré technique, *Exp. Mech.*, Vol. 7, No. 12, pp. 537–540 (1967).

DUNCAN, J. P., Interferometry applied to the study of elastic flexure, *Proc. Inst. Mech. Engrs.*, *London*, Vol. 176, pp. 379–389 (1962).

DUNCAN, J. P., and BROWN, C. J. E., Slope contours in flexed elastic plates by the Saiet–Ikeda technique, *Proc. First Int. Cong. Exp. Mech.*, pp. 149–176 (1963).

DUNCAN, J. P., and MICHEJDA, O., Gridwork rigidity determined by interferometry, *Proc. Inst. Mech. Engrs. London*, Vol. 176, pp. 390–407 (1962).

DUNCAN, J. P., and SABIN, P. G., Determination of curvatures in flexed elastic plates by the Martinelli–Ronchi technique, *Exp. Mech.*, Vol. 3, pp. 285–293 (1963).

DUNCAN, J. P., and SABIN, P. G., An experimental method for recording curvature contours in flexed elastic plates, *Exp. Mech.*, Vol. 5, No. 1, pp. 22–23 (1965).

DURELLI, A. J., Visual representation of the kinematics of the continuum, *Exp. Mech.*, Vol. 6, No. 3, pp. 113–139 (1966).

DURELLI, A. J., and DANIEL, I. M., Structural model analysis by means of moiré fringes, *Proc. ASCE, J. Struct. Div.*, Vol. 86, ST12, pp. 93–102 (1960).

DURELLI, A. J., RILEY, W. F., and CAREY, J. J., Stress distribution on the boundary of a square hole in a large plate during passage of a stress pulse of long duration, Symposium on Photoelasticity, M. M. Frocht (Ed.), Pergamon Press, Oxford, pp. 251–264 (1963).

DURELLI, A. J., and SCIAMMARELLA, C. A., Elastoplastic stress and strain distribution in a finite plate with a circular hole subjected to unidimensional load, *J. Appl. Mech.*, Vol. 30; *Trans. ASME*, Vol. 85, pp. 115–121 (1963).

DURELLI, A. J., SCIAMMARELLA, C. A., and PARKS, V. J., Interpretation of moiré patterns, *Proc. ASCE, J. Struct. Div.*, Vol. 89, EM2, pp. 71–88 (1963).

DURELLI, A. J., and PARKS, V. J., Moiré fringes as parametric curves, *Exp. Mech.*, Vol. 7, No. 3, pp. 97–104 (1967).

DUYSTER, T. H., Scheve aan twee tegenover elkaar gelegen randen oppelegd de overije twee vrij, Institut T.N.O., weer bouwmaterialen and bouwconstructies, Delft, Report 27545 (1957).

DYSON, G., Circular and spiral diffraction gratings, *Proc. Roy. Soc. London*, A, Vol. 248, No. 1282, pp. 93–106 (1958).

EBBENI, G., Développements de la méthode des moirés, final dissertation, Univ. of Brussels (1961).

EBBENI, J., Observation en incidence oblique des phénomènes de moiré par réflexion sur une plaque gauchie: I, *Bull. Acad. Roy. Belg.*, 5 série, Vol. 50, No. 2, pp. 114–124 (1964).

EBBENI, J., Influence des défauts des réseaux et de la diffraction sur la forme et le contraste des franges de moiré, *Bull. Acad. Roy. Belg.*, 5 série, Vol. 50, No. 6, pp. 706–723 (1964).

EBBENI, J., Développements de la méthode des moirés obtenus par réflexion d'un réseau plan sur une surface gauchie: I, *Bull. Acad. Roy. Belg.*, 5 série, Vol. 51, No. 1, pp. 94–115 (1965).

EBBENI, J., Développements de la méthode des moirés obtenus par réflexion d'un réseau plan sur une surface gauchie: II, *Bull. Acad. Roy. Belg.*, 5 série, Vol. 51, No. 2, pp. 207–217 (1965).

EBBENI, J., Etude du phénomène de moiré par reflection d'un réseau plan sur une surface gauchie et de son application en analyse des contraintes et des déformations, *VDI-Berichte*, No. 102, pp. 75–78 (1966).

EVANS, H. E., and DAVIES, R. E., Application of the moiré method to the problem of road pavement analysis, *Strain*, Vol. 1, No. 4, pp. 16–20 (1965).

EVENSEN, D. A., High-speed photographic observation of the buckling of thin cylinders, *Exp. Mech.*, Vol. 4, No. 1, pp. 110–117 (1964).

FAURE, G., Les procédés extensometriques utilisables à chaud, *Revue Fr. de Méc.*, No. 16, pp. 25–34 (1965).

FIDLER, R., and NURSE, P., A method of ruling circular diffraction gratings and their use in the moiré technique of strain analysis, *J. Strain Anal.*, Vol. 1, No. 2, pp. 160–164 (1966).

FIDLER, R., and NURSE, P., Developments in the technique of strain analysis by the moiré method, *VDI-Berichte*, No. 102, pp. 59–64 (1966).

FOUCAULT, L., Mémoire sur la construction des télescopes en verre argenté, *Annls. Obs. Paris*, Vol. 5, No. 197, pp. 197–237 (1859).

FRANCKEN, L., Applications et développements de la méthode des moirés, final dissertation, Univ. of Brussels (1961).

FRAPPIER, E., Etude d'un interféromètre pour la détermination des lignes isopachiques, *Analyse des Contraintes, Mém. GAMAC*, Vol. 2, No. 8, pp. 29–36 (1957).

GATES, J. W., The evaluation of interferograms by displacement and stereoscopic methods, *Brit. J. Appl. Phys.*, Vol. 5, pp. 133–135 (1954).

GEVERS, R., Multiple beam moiré patterns, *Physica Status Solidi*, Vol. 3, No. 12, pp. 2289–2297 (1963).

HALL, R. G. N., and SAYCE, L. A., On the production of diffraction gratings: II, The generation of helical rulings and the preparation of plane gratings therefrom, *Proc. Roy. Soc. London*, 2, Vol. 215, pp. 536–550 (1952).

HARVEY, J., and DUNCAN, J. P., The rigidity of ribreinforced cover plates, *Proc. Inst. Mech. Engrs.*, Vol. 177, No. 5, pp. 115–121 (1963).

HASHIMOTO, H., and UYEDA, R., Detection of dislocations by the moiré pattern in electron micrographs, *Acta Crystallogr.*, London, Vol. 10, pp. 143–148 (1957).

HAZELL, C. R., Experimental investigation of subsequent yield surfaces using the moiré method in plates, Ph.D. thesis, Penn. State Univ. (1965).

HEISE, U., A moiré method for measuring plate curvature, *Exp. Mech.*, Vol. 7, No. 1, pp. 47–48 (1967).

HILLIER, J., New interference phenomena in the electron microscopic images of plate-like crystals, Nat. Bur. Stand., Circ. No. 527, pp. 413–416 (1954).

HOLISTER, G. S., Moiré method of surface strain measurement, *The Engineer*, pp. 149–152 (1967).

HOLISTER, G. S., JONES, W. E. M., and LUXMOORE, A. R., A moiré extenso-meter, *Strain*, Vol. 2, No. 4, pp. 27–33 (1966).

HOLISTER, G. S., and LUXMOORE, A. R., The production of high density moiré grids, *Exp. Mech.*, Vol. 8, No. 5, pp. 210–216 (1968).

IKEDA, K., Soap film technique for solving torsion problems, *Proc. First Jap. Nat. Cong. Appl. Mech.*, Vol. 1, No. 38, pp. 219–224 (1951).

IOSIPESCU, N., Etude expérimentale des plaques planes par la méthode des interférances mécaniques, *Mecanica Aplicata*, Vol. 12, No. 1, pp. 111–137 (1961).

JENKINS, C. J., Diffraction gratings for strain measurement, Univ. of Tasmania, Civil Engrg. Dept., Tech. Report No. 1, pp. 1–12 (1965).

JOHNSON, I., An optical method for studying buckling of plates, *Proc. Soc. Exp. Stress Anal.*, Vol. 16, No. 2, pp. 145–152 (1959).

KACZER, J., and KROUPA, F., The determination of strains by mechanical interference, *Czech. J. Appl. Phys.*, Vol. 1, No. 2, pp. 80–85 (1952).

KOBAYASHI, A. S., BRADLEY, W. B., and SELBY, R. A., Transient analysis in a fracturing epoxy plate with a central notch, *Proc. First Int. Conf. Fracture, Sendai, Japan*, Vol. 3, pp. 1809–1831 (1965).

KOSTAK, B., and POPP, K., Moiré strain gauges, *Strain*, Vol. 2, No. 2, pp. 5–16 (1966).

LANDWEHR, R., and GRABERT, G., Interferenzoptische Versuch zur Platten-biegung, *Ing.-Arch.*, Vol. 18, No. 1, pp. 1–4 (1950).

LAU, E., Beugungserscheinungen an Doppelrastern, *Ann. Phys.*, Vol. 2, No. 7–8, pp. 417–423 (1948).

LAU, E., and MUETZE, K., Ein Dioptriemeter, *Optik*, Vol. 8, pp. 419–425 (1951).

LAU, E., and MUETZE, K., Verwendung von Doppelrastern in der Optik, *Wiss. Annln.*, Vol. 1, pp. 43–56 (1952).

LEES, S., On superposing of two cross-line screens at small angles and the patterns obtained thereby, *Manchr. Phil. Soc. Mem.*, Vol. 63, No. 4, pp. 1–26 (1918–19).

LEHMANN, R., MUETZE, K., VOIGH, H., and WIEMER, A., Das Doppelraster-prinzip als Grundlage für die Entwieklung von Messgeräten mit sehr hoher Empfindlichkeit, *Feingeräte-Tech.*, Vol. 2, No. 4, pp. 153–160 (1953).

LEHMANN, R., and WIEMER, A., Untersuchungen zur Theorie der Doppelraster als Mittel zur Messanzeigen, *Feingeräte-Tech.*, Vol. 2, No. 5, pp. 199–205 (1953).

LIGTENBERG, F. K., Over enn methode om door een eenvonding experiment de momenten in styeve platen te bepalen, *Ingenieur*, Vol. 64, No. 9, pp. 42–46 (1952).

LIGTENBERG, F. K., The moiré method: A new experimental method for the determination of moments in small slab models, *Proc. Soc. Exp. Stress Anal.*, Vol. 12, No. 2, pp. 83–98 (1954).

LIGTENBERG, F. K., and BOUWKAMP, J. G., Experimentele berekening van een scheve plaatbrug, Afd. Wegen Waterbouwkunde, Delft, Technische Hogeschool, Report H–11 (1952).

LIGTENBERG, F. K., and LOOF, H. W., De Spanningstoestand in Dijken, Stevinlaboratorium, Report DV–5 (1957).

LINGE, J. R., Mechanical interference in the measurement of strain, *Aircr. Engng.*, Vol. 29, No. 337, pp. 70–74 (1957).

LOOF, H. W., Onderzoek narr de temperaturspanningen in hét dwarsprofiel van het onderwatergedeelte van de IJ-tunnel te Amsterdam, Stevin-laboratorium, Report TU–13 (1956).

LOOF, H. W., Experimentale onderzoek van temperatuurspanningen, Afd. Wegen Waterbouwkunde, Delft, Technische Hogeschool, Report TS–1 (1957).

LOOF, H. W., and VAN DER SANDE, G. A. F., New fields of application for the moiré method, *Selected Papers of Stress Analysis*, presented at the Institute of Physics, Stress Analysis Group Conference, Delft, 1959, Institute of Physics, England, pp. 20–23 (1961).

LOW, I. A. B., and BRAY, J. W., Strain analysis using moiré fringes, *The Engineer*, pp. 566–569 (1962).

MCLAREN, D. D., The photogrid process for measuring strains caused by underwater explosions, *Proc. Soc. Exp. Stress Anal.*, Vol. 5, No. 2, pp. 115–124 (1948).

MCILRAITH, A. H., A moiré fringe interpolator of high resolution, *J. Sci. Instrum.*, Vol. 41, pp. 34–37 (1964).

MAGUIN, H., *Traité de Microscope Electronique*, Hermans & Co., Paris, Vol. 1, p. 495 (1962).

MENTER, J. W., The direct study by electron microscopy of crystal lattices and their imperfections, *Proc. Roy. Soc. London*, A, Vol. 236, pp. 119–135 (1956).

MENTER, J. W., The electron microscopy of crystal lattices, *Adv. Phys.*, Vol. 7, pp. 299–348 (1958).

MENTER, J. W., Observations on crystal lattices and imperfections by transmission electron microscopes through thin films, *Proc. Fourth Int. Conf. Electron Microsc., Berlin, Sept. 10–17, 1958*, Springer-Verlag, Berlin, Vol. 1, pp. 320–331 (1960).

MERRILL, P. S., Photodot investigation of plastic-strain pattern in flat sheet with a hole, *Exp. Mech.*, Vol. 1, No. 8, pp. 73–80 (1961).

MERTON, T., On the reproduction and ruling of diffraction gratings, *Proc. Roy. Soc. London*, A, Vol. 201, pp. 187–191 (1950).

MERTON, T., Nouvelles méthodes de fabrication des réseaux, *J. Phys. Radium*, Vol. 13, No. 2, pp. 49–53 (1952).

MESMER, G., The interference screen method for isopachic patterns, *Proc. Soc. Exp. Stress Anal.*, Vol. 13, No. 2, pp. 21–26 (1956).

MEYER, M. L., Interpretation of surface-strain measurements in terms of finite homogeneous strains, *Exp. Mech.*, Vol. 3, No. 12, pp. 294–301 (1963).

MIDDLETON, E., Method to increase the accuracy of the moiré method, discussion of a paper by F. P. Chiang, *Proc. ASCE, J. Engng. Mech. Div.*, Vol. 91, EM5, pp. 227–232 (1965).

MIDDLETON, E., A reflection technique for the survey of the deflection of flat plates, *Exp. Mech.*, Vol. 8, No. 2, pp. 56–62 (1968).

MIDDLETON, E., and JENKINS, C. J., A moiré method of strain analysis for student use, *Bull. Mech. Engng. Educ.*, Vol. 5, pp. 257–267 (1966).

MIDDLETON, E., and STEVENSON, L. P., A reflex spectrographic technique for in-plane strain analysis, *Exp. Mech.*, Vol. 8. No. 1, pp. 19–24 (1968).

MILLER, J. A., Improved photogrid techniques for determination of strain over short gauge lengths, *Proc. Soc. Exp. Stress Anal.*, Vol. 10, No. 1, pp. 29–34 (1953).

MIRAU, A., Amplification par trames, *Mésur. Contrôles Ind.*, Vol. 20, No. 212, pp. 23–26 (1955).

MITSUISHI, T., NAGASAKI, H., and UYEDA, R., A new type of interference fringes observed in electron micrograph of crystalline substance, *Proc. Japan Acad.*, Vol. 27, No. 2, pp. 86–87 (1951).

MOORE, A. D., Soap film and sandbed mapper techniques, *J. Appl. Mech.*, Vol. 17, No. 3, pp. 291–298 (1950).

MORSE, S., DURELLI, A. J., and SCIAMMARELLA, C. A., Geometry of moiré fringes in strain analysis, *Proc. ASCE, J. Engng. Mech. Div.*, Vol. 86, EM4, pp. 105–126 (1960).

MOZER, J., BAHCEVANDZIEV, S., JONOSKA, M., and KANICIEVIC, L. J., Moiré patterns of zone plates, *Annu. Fac. Sci. Univ. Skopje, Yougoslavie*, Vol. 15A, pp. 113–147 (1964).

NICKOLA, W. E., The dynamic response of thin membranes by the moiré method, *Exp. Mech.*, Vol. 6, No. 12, pp. 593–601 (1966).

NICKOLA, W. E., CONWAY, H. D., and FARNHAM, K. A., Moiré study of anticlastic deformations of strips with tapered edges, *Exp. Mech.*, Vol. 7, No. 4, pp. 168–175 (1967).

NISHIJIMA, Y., and OSTER, G., Moiré patterns: Their application to refractive index and refractive index gradient measurements, *J. Opt. Soc. Am.*, Vol. 54, No. 1, pp. 1–5 (1964).

NISIDA, M., and SAITO, H., A new interferometric method of two-dimensional stress analysis, *Exp. Mech.*, Vol. 4, No. 2, pp. 366–376 (1964).

NISIDA, M., and SAITO, H., Stress distributions in a semi-infinite plate due to a pin determined by interferometric method, *Exp. Mech.*, Vol. 6, No. 5, pp. 273–279 (1966).

O'HAVEN, C. P., and HARDING, J. F., Studies of plastic flow problems by photogrid methods, *Proc. Soc. Exp. Stress Anal.*, Vol. 2, No. 2, pp. 59–70 (1945).

OPPEL, G. U., and HILL, P. W., Strain measurements at the root of cracks and notches, *Exp. Mech.*, Vol. 4, No. 2, pp. 206–211 (1964).

OSGERBY, C., Application of the moiré method for use with cylindrical surfaces, *Exp. Mech.*, Vol. 7, No. 7, pp. 313–320 (1967).

OSTER, G., Representation and solution of optical problems by moiré patterns, Symposium on quasi-optics, Interscience Publ., J. Wiley, New York, pp. 59–68 (1964).

OSTER, G., Optical art, *Appl. Opt.*, Vol. 4, pp. 1359–1371 (1965).

OSTER, G., Moiré optics: a bibliography, *J. Opt. Soc. Am.*, Vol. 55, No. 10, pp. 1329–1330 (1965).

OSTER, G., and NISHIJIMA, Y., Moiré patterns, *Scient. Am.*, pp. 54–63 (1963).

OSTER, G., WASSERMAN, M., and ZWERLING, C., Theoretical interpretation of moiré patterns, *J. Opt. Soc. Am.*, Vol. 44, No. 2, pp. 169–175 (1964).

OSTERBERG, H., An interferometer method of studying the vibrations of an oscillating quartz plate, *J. Opt. Soc. Am.*, Vol. 22, pp. 19–35 (1932).

PALMER, P. J., The bending stresses in cantilever plates by moiré fringes, *Aircr. Engng.*, Vol. 29, No. 346, pp. 377–380 (1957).

PARKS, V. G., Moiré grid-analyser method for strain analysis (discussion), *Exp. Mech.*, Vol. 6, No. 5, pp. 287–288 (1966).

PARKS, V. J., and DURELLI, A. J., Various forms of the strain-displacement relations applied to experimental stress analysis, *Exp. Mech.*, Vol. 4, pp. 33–47 (1964).

PARKS, V. J., and DURELLI, A. J., Moiré patterns of partial derivatives of displacement components, *J. Appl. Mech.*, Vol. 33, Series E, No. 4, pp. 901–906 (1966).

PASHLEY, D. W., MENTER, J. M., and BASSET, G. A., Observation of dislocations in metals by means of moiré patterns on electron micrographs, *Nature*, Vol. 179, No. 4563, pp. 752–755 (1957).

PEERS, N., Utilisation d'un procédé de moirure en photoélasticimétrie, final dissertation, Univ. of Brussels (1957).

PIRARD, A., Consideration sur la méthode du moiré en photoélasticité, *Analyse des Contraintes, Mém. GAMAC*, Vol. 5, No. 2, pp. 1–24 (1960).

PIRARD, A., Resolution photoélastique des systèmes hyperstatiques, *Révue Fr. Méc.*, Vol. 2, No. 4, pp. 129–138 (1962).

POST, D., Photoelastic stress analysis for an edge-crack in a tensile field, *Proc. Soc. Exp. Stress Anal.*, Vol. 12, No. 1, pp. 99–116 (1954).

POST, D., A new photoelastic interferometer suitable for static and dynamic measurements, *Proc. Soc. Exp. Stress Anal.*, Vol. 12, No. 1, pp. 191–202 (1954).

POST, D., Photoelastic evaluation of individual principal stresses by large field absolute retardation measurements, *Proc. Soc. Exp. Stress Anal.*, Vol. 13, No. 2, pp. 119–132 (1955).

POST, D., The moiré grid-analyser method for strain analysis, *Exp. Mech.*, Vol. 5, No. 11, pp. 368–377 (1965).

POST, D., Sharpening and multiplication of moiré fringes, *Exp. Mech.*, Vol. 7, No. 4, pp. 154–159 (1967).

POST, D., New optical methods of moiré fringe multiplication, *Exp. Mech.*, Vol. 8, No. 2, pp. 63–68 (1968).

RAMAN, C. V., and DATTA, S. K., On Brewster's bands: I, *Trans. Opt. Soc. London*, Vol. 27, No. 7, pp. 51–55 (1926).

RANG, O., Ferninterferenzen von electronenwellen, *Z. Phys.*, Vol. 136, pp. 465–479 (1953).

RAYLEIGH, LORD, On the manufacture and theory of diffraction gratings, *Phil. Mag.*, Vol. 47, No. 310, pp. 81–93; No. 311, pp. 193–205 (1874).

RAYLEIGH, LORD, *Scientific Papers of Lord Rayleigh*, Cambridge Univ. Press, Vol. 1, p. 92 (1874).

RICHARDS, T. H., Analogy between the slow motion of a viscous fluid and the extension and flexure of plates; a geometric demonstration by means of moiré fringes, *Brit. J. Appl. Phys.*, Vol. 11, pp. 244–254 (1960).

RIEDER, G., and RITTER, R., Krümmungsmessung an belasteten Platten nach dem Ligtenbergschen Moiré-Verfahren, *Forsch. Ing.-Wes.*, Vol. 31, No. 2, pp. 33–44 (1965).

RIGHI, A., Sui fenomeni che si producono colla sovrapposizione di due reticoli e sopra alcune lora applicazioni: I, *Nuovo Cim.*, Vol. 21, pp. 203–227 (1887).

RIGHI, A., Sui fenomeni che si producono colla sovrapposizione di due reticoli e sopra alcune lora applicazioni: II, *Nuovo Cim.*, Vol. 22, pp. 10–43 (1888).

RILEY, W. F., and DURELLI, A. J., Application of moiré methods to the determination of transient stress and strain distribution, *J. Appl. Mech.*, Vol. 29; *Trans. ASME*, Vol. 84, pp. 23–29 (1962).

ROGERS, G. L., A simple method of calculating moiré patterns, *Proc. Phys. Soc.*, Vol. 73, pp. 142–144 (1959).

ROGOZINSKI, M., An attempt to establish the theoretical foundations of the moiré method of strain and stress analysis, *Arch. Appl. Mech.* (Polokieak Nauk.), Vol. 9, No. 2, pp. 191–206 (1957).

RONCHI, V., La prova dei sistemi ottici, *Attual. Scient.*, No. 37 (1925).

RONCHI, V., Sur la nature interferentielle des franges d'ombre dans l'essai des systèmes optiques, *Revue Opt.*, Vol. 5, No. 11, pp. 431–437 (1926).

RONCHI, V., Das Okularinterferometer und das Objektivinterferometer bei Auflösung der Doppelsterne, *Z. Phys.*, Vol. 37, No. 10, pp. 732–757 (1926).

RONCHI, V., *Corso di ottica technica: II*, Associazione Ottica Italiana, Firenze, Italy (1954).

RONCHI, V., Forty years of history of a grating interferometer, *Appl. Opt.*, Vol. 3, No. 4, pp. 437–451 (1964).

ROSS, B. E., SCIAMMARELLA, C. A., and STURGEON, D., Basic optical law in the interpretation of moiré patterns applied to the analysis of strains: 2, *Exp. Mech.*, Vol. 5, No. 6, pp. 161–166 (1965).

SALET, G., Nouvelle méthode de mise en oeuvre de l'analogie de la membrane pour l'étude de la torsion des poutres cylindriques, *Tech. Mar. Aeronaut.*, Vol. 43, pp. 107–119 (1939).

SAMPSON, R. C., and CAMPBELL, D. M., The Gridshift technique for moiré analysis of strain in solid propellants, *Exp. Mech.*, Vol. 7, No. 11, pp. 449–457 (1967).

SANDERS, J. B., Precise topography of optical surfaces, *J. Res.*, Vol. 47, No. 3, p. 148 (1951).

SATTERLY, J., and STUCKEY, E. L., The shape of the profile of a liquid film draining on a vertical wetted plate studied by interference fringes and by direct photography; also an investigation of the so-called plate ridge by the use of interference fringes, *Trans. Roy. Soc. Canada*, Vol. 3, pp. 131–161 (1932).

SAYCE, L. A., The preparation and industrial applications of diffraction gratings, *Trans. Soc. Instrum. Technol.*, Vol. 9, pp. 139–143 (1957).

SAYCE, L. A., and FRANKS, A., N.P.L. gratings for X-ray spectroscopy, *Proc. Roy. Soc. London*, Vol. 282, pp. 353–357 (1964).

SCIAMMARELLA, C. A., Basic optical law in the interpretation of moiré patterns applied to the analysis of strains: I, *Exp. Mech.*, Vol. 5, pp. 154–160 (1965).

SCIAMMARELLA, C. A., Techniques of fringe interpolation in moiré patterns, *Proc. Second SESA Int. Congr. Exp. Mech.*, B. E. Rossi (Ed.), pp. 62–73 (1966), and *Exp. Mech.*, Vol. 7, No. 11, pp. 19A–30A (1967).

SCIAMMARELLA, C. A., Gap effect in moiré fringes observed with coherent monochromatic collimated light, to be published in *Exp. Mech.*

SCIAMMARELLA, C. A., and CHIANG, FU-PEN, The moiré method applied to three-dimensional elastic problems, *Exp. Mech.*, Vol. 4, No. 2, pp. 313–319 (1964).

SCIAMMARELLA, C. A., and DODDINGTON, C. W., Effect of photographic-film nonlinearities on the processing of moiré fringe data, *Exp. Mech.*, Vol. 7, No. 9, pp. 398–402 (1967).

SCIAMMARELLA, C. A., and DURELLI, A. J., Moiré fringes as a means of analysing strains, *Proc. ASCE, J. Engng. Mech. Div.*, Vol. 87, EM1, pp. 55–74 (1961).

SCIAMMARELLA, C. A., and ROSS, B. E., Thermal stresses in cylinders by the moiré method, *Proc. Soc. Exp. Stress Anal.*, Vol. 21, No. 2, pp. 289–296 (1964).

SCIAMMARELLA, C. A., and STURGEON, D. L., Substantial improvements in the processing of moiré data by optical and digital filtering, *VDI-Berichte*, No. 102, pp. 71–74 (1966).

SCIAMMARELLA, C. A., and STURGEON, D. L., Thermal stresses at high temperatures in strainless-steel rings by the moiré method, *Exp. Mech.*, Vol. 6, No. 5, pp. 235–243 (1966).

SCIAMMARELLA, C. A., and STURGEON, D. L., Digital-filtering techniques applied to the interpolation of moiré fringes data, *Exp. Mech.*, Vol. 7, No. 11, pp. 468–475 (1967).

SHEPHERD, R. R., and McD. WENSLEY, L., The moiré fringe method of displacement measurement applied to indirect structural-model analysis, *Exp. Mech.*, Vol. 5, No. 6, pp. 167–176 (1965).

SINCLAIR, D., A new optical method for the determination of the principal stress sum, *Proc. Tenth Semi-Annual Eastern Photoelasticity Conf.* (1939).

STURGEON, D. L., Analysis and synthesis of a moiré photo-optical system, *Exp. Mech.*, Vol. 7, No. 8, pp. 346–352 (1967).

SZMODITS, K., Zur theorie des moiré verfahrens, *Proc. Symp. Shell Res. (Delft)*, North Holland Publ. Co., Amsterdam, pp. 208–216 (1961).

TANAKA, K., and NAKASHIMA, M., Plastic deformation of a plate containing a circular hole, *Proc. Fifth Japan Cong. Test. Mater.*, Japan Soc. Testing Materials, pp. 82–85 (1962).

TANAKA, K., and NAKASHIMA, M., Application of the photographing technique to the measurement of the strain produced at high temperature, *Proc. Sixth Japan. Cong. Test. Mater.*, Japan Soc. Testing Materials, pp. 79–81 (1963).

THEOCARIS, P. S., On the variation of contraction ratio in the elastoplastic domain of metals, *Proc. Nat. Acad. Sci. Athens*, Vol. 36, pp. 238–247 (1961).

THEOCARIS, P. S., The variation of Poisson's ratio in the elastoplastic domain of metals, *Proc. Am. Soc. Test. Mater.*, Vol. 61, pp. 838–846 (1961).

THEOCARIS, P. S., Moiré fringes: a powerful measuring device, *Appl. Mech. Rev.*, Vol. 15, No. 5, pp. 333–339 (1962).

THEOCARIS, P. S., On a simple interferometric method for the separation of principal stresses in plane elasticity problems, *Proc. Nat. Acad. Sci. Athens*, Vol. 37, pp. 273–284 (1962).

THEOCARIS, P. S., Moiré method in plates, *Proc. Int. Assoc. Shell Struct.* (symposium), Warsaw, North Holland Publ. Co., Amsterdam, pp. 877–889 (1963).

THEOCARIS, P. S., Diffused light interferometry for measurement of isopachics, *J. Mech. Phys. Solids*, Vol. 11, No. 3, pp. 181–195 (1963).

THEOCARIS, P. S., Moiré fringes: a powerful measuring device (in Greek), *Tech. Ann.*, Vol. 44, No. 3, pp. 315–347 (1963).

THEOCARIS, P. S., The moiré method in thermal fields, *Exp. Mech.*, Vol. 4, No. 8, pp. 223–231 (1964).

THEOCARIS, P. S., Moiré patterns of isopachics, *J. Sci. Instrum.*, Vol. 41, pp. 133–138 (1964).

THEOCARIS, P. S., Isopachic patterns by the moiré method, *Exp. Mech.*, Vol. 4, No. 6, pp. 153–159 (1964).

THEOCARIS, P. S., A review of the rheo-optical properties of linear high polymers, *Exp. Mech.*, Vol. 5, No. 4, pp. 105–114 (1965).

THEOCARIS, P. S., A method for measuring thickness changes in plane transparent models, *Exp. Mech.* (discussion), Vol. 5, No. 11, pp. 365–367 (1965).

THEOCARIS, P. S., Moiré patterns of slope contours in flexed plates, *Proc. Second SESA Int. Congr. Exp. Mech.*, B. E. Rossi (Ed.), pp. 56–61 (1966), and *Exp. Mech.*, Vol. 6, No. 4, pp. 212–217 (1966).

THEOCARIS, P. S., Moiré analysis of cylindrical surfaces; Experimentelle Spannungsanalyse, *VDI-Berichte*, No. 102, pp. 79–83 (1966).

THEOCARIS, P. S., Moiré fringes: a powerful measuring device, *Applied Mechanics Surveys*, Spartan Books Inc., pp. 613–626 (1966).

THEOCARIS, P. S., Flexural vibrations of plates by the moiré method, *Brit. J. Appl. Phys.*, Vol. 18, No. 4, pp. 513–519 (1967).

THEOCARIS, P. S., Moiré topography of curved surfaces, *Exp. Mech.*, Vol. 7, No. 7, pp. 289–296 (1967).

THEOCARIS, P. S., The dynamic response of thin membranes by the moiré method (discussion), *Exp. Mech.*, to be published.

THEOCARIS, P. S., Fractional moiré fringes by isodensitracing, *Quart. Sci. Rev.*, Vol. 36, Nos. 3–4, pp. 626–631 (1967).

THEOCARIS, P. S., Thermal stress at high temperatures in strainless-steel rings by the moiré method (discussion), *Exp. Mech.*, Vol. 8, No. 6, pp. 270–278 (1968).

THEOCARIS, P. S., Optical nodal pattern analysis, *Exp. Mech.*, Vol. 8, No. 5, pp. 237–240 (1968).

THEOCARIS, P. S., A moiré method for measuring plate curvature (discussion), *Exp. Mech.*, Vol. 8, No. 8, to be published.

THEOCARIS, P. S., Radial gratings as moiré gauges, *J. Sci. Instrum.* (*J. Phys.* E), Ser. 2, Vol. 1, No. 6, pp. 613–618 (1968).

THEOCARIS, P. S., Moiré patterns formed by inclined gratings, *Brit. J. Appl. Phys.* (*J. Phys.* D), Ser. 2, Vol. 1, No. 7, pp. 420–428 (1968).

THEOCARIS, P. S., Curvature distribution obtained by frequency modulated amplitude gratings, *J. Sci. Instrum.* (*J. Phys.* E), Ser. 2, Vol. 1, No. 6, pp. 619–622 (1968).

THEOCARIS, P. S., Direct tracing of isoentatics by moiré patterns, *Material-prüfung*, Vol. 10, No. 5, pp. 155–158 (1968).

THEOCARIS, P. S., Non-uniform plastic fields in pure compression; Mechanics of the solid state, J. Marin's anniversary volume, F. P. J. Rimrott (Ed.), pp. 240–258 (1968).

THEOCARIS, P. S., DAVIDS, N., GILLICH, W., and CALVIT, H. H., The moiré method for the study of explosions, *Exp. Mech.*, Vol. 7, No. 5, pp. 202–210 (1967).

THEOCARIS, P. S., GILLICH, W., and MARKETOS, E., Shock wave propagation in perspex spheres, *Int. J. Mech. Sci.*, Vol. 8, No. 6, pp. 739–749 (1966).

THEOCARIS, P. S., and HADJIJOSEPH, C., Transient lateral contraction ratio of polymers in creep and relaxation, *Kolloid-Z.*, Vol. 202, No. 2, pp. 133–139 (1965).

THEOCARIS, P. S., and HAZELL, C. R., Experimental investigation of subsequent yield surfaces using the moiré method, *J. Mech. Phys. Solids*, Vol. 13, No. 6, pp. 281–294 (1965).

THEOCARIS, P. S., and KORONEOS, E., Stress-strain and contraction ratio curves for polycrystalline steel, *Phil. Mag.*, Vol. 8, No. 95, pp. 1871–1893 (1963).

THEOCARIS, P. S., and KORONEOS, E., The variation of the lateral contraction ratio of polycrystalline steel at higher temperatures, *Proc. Am. Soc. Test. Mater.*, Vol. 64, pp. 747–764 (1964).

THEOCARIS, P. S., and KOUTSAMBESSIS, A., The slit-source and grating method in plane stress problems, *Strain*, Vol. 4, No. 1, pp. 10–15 (1968).

THEOCARIS, P. S., and KOUTSAMBESSIS, A., Slope measurement by means of moiré fringes, *J. Sci. Instrum.*, Vol. 42, No. 8, pp. 607–610 (1965).

THEOCARIS, P. S., and KOUTSAMBESSIS, A., The multisource and the slit-source and grating methods for recording slope contours in plates, *Acta Technika Czech.*, Vol. 18, No. 6, pp. 176–184 (1967).

THEOCARIS, P. S., and KOUTSAMBESSIS, A., Surface topography by multisource moiré patterns, *Exp. Mech.*, Vol. 8, No. 2, pp. 82–87 (1968).

THEOCARIS, P. S., and KUO, H. H., The moiré method of zonal and line gratings, *Exp. Mech.*, Vol. 5, No. 8, pp. 267–272 (1965).

THEOCARIS, P. S., and KUO, H. H., Interference patterns of zonal gratings: application to rheology of polymers, *Z. Angew. Math. Phys.*, Vol. 17, No. 1, pp. 90–107 (1966).

THEOCARIS, P. S., and MARKETOS, E., The formation of necking in polycrystalline steel, *Proc. First Int. Conf. Fracture, Sendai, Japan 1965*, Vol. 3, pp. 1781–1807 (1965); see also *Acta Mech.*, Vol. 3, No. 2, pp. 103–122 (1967).

THIRY, R., and DANTU, P., Exposé général d'une méthode expérimentale de détermination directe des déformations dans un solide, *Proc. Ninth Cong. Int. Mech. Appl.*, *Univ. Brux.*, Vol. 8, pp. 490–499 (1957).

TOLLENAAR, D., Moiré interferentieverschijnselen bij rasterdruk, Amsterdam, Institut voor Graphische Technik (1945).

VAFIADAKIS, A. P., Two moiré instruments, *J. Sci. Instrum.*, Vol. 44, No. 12, pp. 1008–1010 (1967).

VAFIADAKIS, A. P., and LAMBLE, J. H., The application of moiré rotational mismatch techniques to strain analysis, *J. Strain Anal.*, Vol. 2, No. 2, pp. 99–108 (1967).

VAN DER SANDE, G. A. F., Het moiré-modelonderzoek also doeltreffende methode voor het bepalen van de momenten in vlakke platen, *Ingenieur*, Vol. 68, No. 13, pp. 017–022 (1956).

VAN OSS, C. J., The use of gratings producing moiré patterns for measuring refractive index gradients, *J. Sci. Instrum.*, Vol. 41, pp. 227–228 (1964).

VANDIEGHEM, L., Birefringence et deformations, final dissertation, Univ. of Brussels (1961).

VARGADY, L. O., Moiré fringes as visual position indicators, *Appl. Opt.*, Vol. 3, No. 4, pp. 537–542 (1964).

VINCKIER, A., and DECHAENE, R., Use of the moiré effect to measure plastic strains, *Trans. ASME*, Vol. 82D; *J. Bas. Engng.*, Vol. 2, pp. 426–434 (1960).

VREEDENBURGH, C. G. J., and VAN WIJNGAARDEN, H., New progress in our knowledge about the moment distribution in flat slabs by means of the moiré method, *Proc. Soc. Exp. Stress Anal.*, Vol. 12, No. 2, pp. 99–114 (1954).

WASIL, B. A., and MERCHANT, D. C., Plate measurement by photogrammetric methods, *Exp. Mech.*, Vol. 4, No. 2, pp. 77–83 (1964).

WASIL, B. A., MERCHANT, D. C., and DEL VECCHIO, J. J., Photogrammetric measurements of dynamic displacement, *Exp. Mech.*, Vol. 5, No. 10, pp. 332–339 (1965).

WELLER, R., and SHEPARD, B. M., Displacement measurement by mechanical interferometry, *Proc. Soc. Exp. Stress Anal.*, Vol. 6, No. 1, pp. 35–38 (1948).

WHITE, E. K., The application of the moiré method to plane stress problems, M.Sc. thesis Research Report, Div. Structural Engng. and Structural Mech., Univ. of Calif., Berkeley (1959).

WIEMER, A., LEHMANN, R., and VOIGT, H., Neuer statischer Dehnungsmesser mit 20 mm. Messlänge, *Feingeräte-Techn.*, Vol. 3, No. 4, pp. 161–163 (1954).

ZANDMAN, F., HOLISTER, G. S., and BRCIC, V., The influence of grid geometry on moiré fringe properties, *J. Strain Anal.*, Vol. 1, No. 1, pp. 1–10 (1965).

ZIENKIEWICZ, D. C., and CRUZ, C., The use of the slab analogy in the determination of thermal stresses, *Int. J. Mech. Sci.*, Vol. 4, pp. 285–296 (1962).

Author Index

417

Subject Index

421